T0243047

CAMBRIDGE LIBRARY COLLECTION

Books of enduring scholarly value

Botany and Horticulture

Until the nineteenth century, the investigation of natural phenomena, plants and animals was considered either the preserve of elite scholars or a pastime for the leisured upper classes. As increasing academic rigour and systematisation was brought to the study of 'natural history', its subdisciplines were adopted into university curricula, and learned societies (such as the Royal Horticultural Society, founded in 1804) were established to support research in these areas. A related development was strong enthusiasm for exotic garden plants, which resulted in plant collecting expeditions to every corner of the globe, sometimes with tragic consequences. This series includes accounts of some of those expeditions, detailed reference works on the flora of different regions, and practical advice for amateur and professional gardeners.

London Parks and Gardens

Brought up among the extensive grounds of her family home at Didlington Hall in Norfolk, Alicia Amherst (1865–1941) was a keen gardener from an early age. Especially interested in socially beneficial gardening, she sat on the board of the Chelsea Physic Garden from 1900, encouraged the growing of smoke-resistant flowers in poor urban areas, and promoted the greater use of allotments and school gardens during the First World War. Long regarded as a significant work for its thorough yet accessible approach, this well-researched historical and horticultural survey first appeared in 1907 under her married name of the Honourable Mrs Evelyn Cecil. Beautifully illustrated throughout, it covers London's royal and other parks as well as less obvious green spaces such as squares, burial grounds, and Inns of Court. A map and plant lists are also included. Amherst's *History of Gardening in England* (1895) is also reissued in this series.

Cambridge University Press has long been a pioneer in the reissuing of out-of-print titles from its own backlist, producing digital reprints of books that are still sought after by scholars and students but could not be reprinted economically using traditional technology. The Cambridge Library Collection extends this activity to a wider range of books which are still of importance to researchers and professionals, either for the source material they contain, or as landmarks in the history of their academic discipline.

Drawing from the world-renowned collections in the Cambridge University Library and other partner libraries, and guided by the advice of experts in each subject area, Cambridge University Press is using state-of-the-art scanning machines in its own Printing House to capture the content of each book selected for inclusion. The files are processed to give a consistently clear, crisp image, and the books finished to the high quality standard for which the Press is recognised around the world. The latest print-on-demand technology ensures that the books will remain available indefinitely, and that orders for single or multiple copies can quickly be supplied.

The Cambridge Library Collection brings back to life books of enduring scholarly value (including out-of-copyright works originally issued by other publishers) across a wide range of disciplines in the humanities and social sciences and in science and technology.

London Parks
and Gardens

Alicia Amherst

CAMBRIDGE
UNIVERSITY PRESS

CAMBRIDGE
UNIVERSITY PRESS

University Printing House, Cambridge, CB2 8BS, United Kingdom

Cambridge University Press is part of the University of Cambridge.
It furthers the University's mission by disseminating knowledge in the pursuit of
education, learning and research at the highest international levels of excellence.

www.cambridge.org
Information on this title: www.cambridge.org/9781108075992

© in this compilation Cambridge University Press 2015

This edition first published 1907
This digitally printed version 2015

ISBN 978-1-108-07599-2 Paperback

LONDON PARKS AND GARDENS

ST. KATHARINE'S LODGE, REGENT'S PARK

LONDON PARKS AND GARDENS

BY

THE HON^BLE MRS. EVELYN CECIL

(ALICIA AMHERST)

CITIZEN AND GARDENER OF LONDON

AUTHOR OF "A HISTORY OF GARDENING IN ENGLAND"
"CHILDREN'S GARDENS," ETC.

WITH ILLUSTRATIONS BY

LADY VICTORIA MANNERS

" Reade the whole and then judge"
JOHN CHRISTOPHERSON,
Bishop of Chichester, 1554

LONDON
ARCHIBALD CONSTABLE & CO. LTD.
1907

Printed by BALLANTYNE, HANSON & CO.
At the Ballantyne Press, Edinburgh

PREFACE

In spite of the abundance of books on London, not one exists which tells the story of the Parks and Gardens as a whole. Some of the Royal Parks have been dealt with, and most of the Municipal Parks, but in separate works. When Squares are touched on, in guide-books, or in volumes to themselves, the Gardens are for the most part left alone, and gossip of the inhabitants forms the centre of the narrative. This is the case also with public buildings and private houses which have gardens attached to them. To give a sketch of the history of the more important Parks and Gardens, and to point out any features of horticultural interest, is the object of the following pages. London is such a wide word, and means such a different area at various periods, that it has been necessary to make some hard and fast rule to define the scope of this work. I have, therefore, decided to keep strictly to the limits of the County of London within the official boundaries of the London County Council at the present time.

I would express my thanks to the authorities of the

Parks, both Royal and Municipal, for their courtesy in affording me information, and to many friends who have facilitated my search in historical and private gardens. I am also extremely grateful to my friend, Miss Margaret MacArthur, who has assisted me in the tedious task of correcting proofs. The lists of trees and shrubs, and of plants in the beds in Hyde Park, were kindly drawn up for me by the Park Superintendent, the late Mr. Jordan, with the consent of H.M. Office of Works.

ALICIA M. CECIL.

10 Eaton Place,
  August 1907.

CONTENTS

CONTENTS

LIST OF ILLUSTRATIONS

COLOURED PLATES

X LIST OF ILLUSTRATIONS

IN THE TEXT

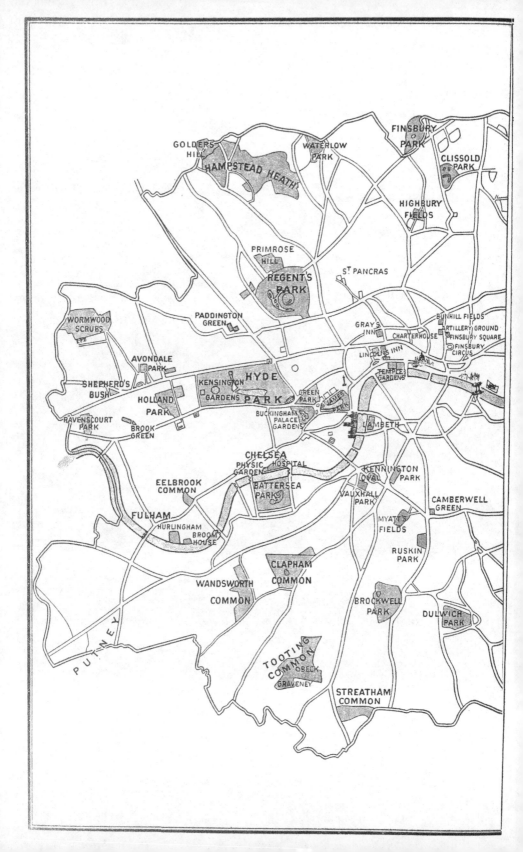

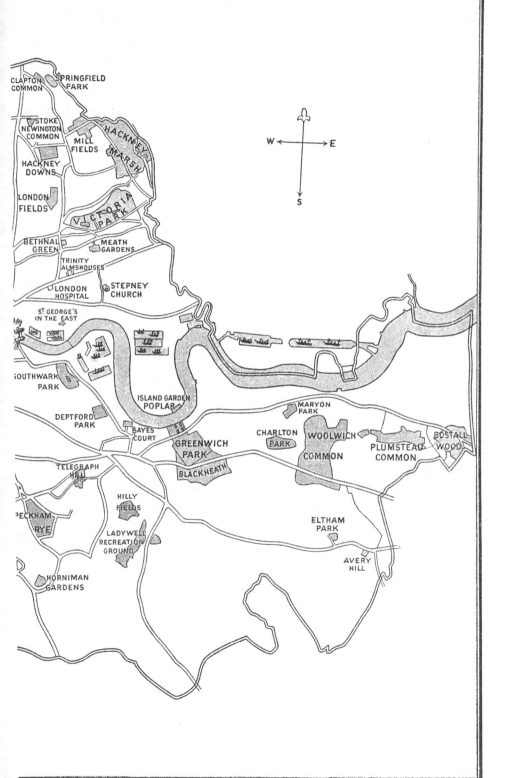

London Parks & Gardens

CHAPTER I

INTRODUCTORY

London, thou art the Flour of cities all.
—WILLIAM DUNBAR, 1465–1530.

ONDON has a peculiar fascination of its own, and to a vast number of English-speaking people all over the world it appeals with irresistible force. So much has been said and written about it that the theme might seem to be worn out, yet there are still fresh aspects to present, still hidden charms to discover, still deep problems to solve. The huge, unwieldly mass, which cannot be managed or legislated for as other towns, but has to be treated as a county, enfolds within its area all the phases of human life. It embraces every gradation from wealth to poverty, from the millionaire to the pauper alien. The collection of buildings which together make London are a most singular assortment of innumerable variations between beauty and ugliness, between palaces and works of art and hovels of sordid and unlovely squalor.

A

An Englishman must be almost without soul who can stand for the first time unmoved within the precincts of Westminster Abbey or look without satisfaction at the faultless proportions of St. Paul's. The sense of possession, the pride of inheritance, are the uppermost feelings in his mind.   But he who loves not only London itself with a patriotic veneration, but also his fellow-men, will not rest with the inspection of the beautiful. He will journey eastward into the heart of the mighty city, and see its seething millions at work, its dismal poverty, its relentless hardness. The responsibility of heirship comes over him, the sadness, the pathos, the evil of it all depresses him, the hopelessness of the contrast overpowers him; but apart from all ideas of social reform, from legislative action or philanthropic theories, there is one thin line of colour running through the gloomy picture. The parks and gardens of London form bright spots in the landscape. They are beyond the pale of controversy; they appeal to all sections of the community, to the workers as well as to the idlers, to the rich as well as to the poor, to the thoughtful as well as to the careless. From the utilitarian point of view they are essential. They bring new supplies of oxygen, and allow the freer circulation of health-giving fresh air. They are not less useful as places of exercise and recreation. They waft a breath of nature where it is most needed, and the part they play in brightening the lives of countless thousands cannot be over-estimated.

The parks and gardens of London have a past full of historical associations, and at the present time their full importance is slowly being realised. Much has been done to improve and beautify them, but much remains to be achieved in that direction before their capabilities will have been thoroughly developed. The opportunity is great, and if only the best use can be made of it London Parks could be the most beautiful as well as the most useful in the world. It is impossible to praise or criticise them collectively, as they have different origins, are administered by separate bodies, and have distinct functions to perform. It cannot be denied that the laying out in some and the planting in other cases could be improved. Plans could be carried out with more taste than is sometimes shown, and new ideas be encouraged, but on the whole there is so much that is excellent and well done that there is a great deal to be proud of.

The various open spaces in London can easily be grouped into classes. First there are the Royal Parks, with a history and management of their own; then there are all the Parks either created or kept up by the London County Council, and most of the commons and other large open spaces are in their jurisdiction also, though a few parks and recreation grounds are under the borough councils. Municipal bodies for the most part take charge of all the disused burial grounds converted into gardens, though some are maintained by the parish or the rector. Then there is another class of garden which must be included, namely, all the squares of London, as, although few are open to the public, they form no insignificant proportion of the unbuilt area.

All through London there are survivals of old gardens, which are still either quiet and concealed, or thrown open to the public. Such are the grounds of the Charterhouse, of Chelsea Hospital, or of the Foundling

Hospital, and of other old-world haunts of peace.   The rarest thing in London are the private gardens, yet they too go to make up the aggregate lungs of the city. Out of a total of upwards of 75,000 acres there are in round numbers some 6000 acres of parks, commons, squares, and open spaces in London : of these a little over 4000 acres are in the hands of the London County Council.   Besides this it administers nearly 900 acres outside the county.   The City of London owns large forest tracts, commons, and parks beyond the limit of the County of London—Epping, Burnham Beeches, Highgate Wood, and parks in West Ham, Kilburn, &c.—altogether nearly 6500 acres.

London is such a wide word, it is difficult to set a limit, and to decide what open spaces actually belong to London.   As the town stretches away into the country, it is impossible to see the boundaries of London. The line must be drawn near where the chimney-pots become incessant, and the stems of the trees become black.   But the degree of blackness, dirt, and density is impossible to decide ; so a prosaic, matter-of-fact, but necessary rule has been adhered to in the following pages, of keeping as strictly as possible to the actual defined limits of the County of London.   Therefore all the parks owned by the City Corporation or London County Council outside this limit have not been dealt with, and such places as Chiswick, Kew, Richmond, or Gunnersbury have been omitted.

To get to some of these places involves a considerable journey.   Many of the outlying parks have to be reached by train, or by a very long drive, or tram ride. From Hyde Park Corner, for instance, to Bostall Wood or Avery Hill is a long expedition.   To the fortunate

few who possess motor cars the distances are trifling, but
the vast majority of people must exercise considerable
ingenuity, and possess a good bump of locality, if they
wish to visit all London's open spaces. A knowledge
of the distant places, the names of which are inscribed
in large letters on every omnibus, is necessary. The
Royal Oak, Elephant and Castle, or Angel, are but
starting-places for the more distant routes, although
they form the goal of green, red, or blue 'busses. The
electric trams of South London have made the approach
to Dulwich, Peckham, Greenwich, and many other parks
much more simple, and motor 'busses rattle along close to
even the distant Golder's Hill or Highbury Fields. With
a railway time-table, a good eye for colour in selecting
the right omnibus, and a knowledge of the points of the
compass, every green patch in London can be reached
with ease, even by those whose purses are not long
enough to let them indulge in motors, or whose nerves
are not steady enough to let them venture on bicycles.

Each park forms the central point of some large
district, and they are not dependent on the casual visitor
for appreciation. Every single green spot, on a fine
Saturday throughout the year, is peopled with a crowd
from the neighbourhood, and on every day in the year,
winter as well as summer, almost every open space has
a ceaseless throng of comers and goers.

What is the cost of maintenance of these parks is a
question that will naturally occur; and the answer in
many cases is easy to find, as the statistics of both the
London County Council Parks, published in their hand-
book, and those of the Royal Parks, which are submitted
to Parliament every year, are accessible. The following
extracts may, however, be useful. In looking at the

two sets of figures, of course the acreage must be borne in mind, and the great expense of police in the Royal Parks, amounting to £8782 for Hyde Park alone, must be deducted before any fair comparison can be made, even when results are not considered.

		1907–8.					1906–7.
	Acres.	Wages and Salaries.	Police, Park-keepers.	New Works and Alterations.	Maintenance.	Total.	Total.
		£	£	£	£	£	£
1. Greenwich .	185	225	1,090	175	3,737	5,319	4,554
2. { Hyde Park, St. James's, Green Park } .	509½	724	12,153	4,965	50,886	69,269	48,835
3. Kensington Gardens	274	138	1,590	50	5,831	7,730	7,804
4. { Regent's Park and Primrose Hill } .	472½	290	2,171	300	11,417	14,542	13,329

Taken from the Estimates for 1907–8.

	Acres.	Net Aggregate Capital Expenditure.	Average Cost of Maintenance.	Number of Staff.
		£	£	
Battersea	199	21,042	10,897	92
Brockwell	127¼	114,322	4,493	34
Dulwich	72	45,510	3,330	28
Finsbury	115	137,934	7,649	52
Victoria	217	38,430	12,099	107
Waterlow	26	11,178	2,658	24

Taken from L.C.C. Handbook No. 1009, 1906.

London has always been a city of gardens, and although much boast is made of the newly-acquired open spaces, a wail for those destroyed would have equal justification. It is very terrible that everything in life

has to be learnt by slow and hard lessons, dearly purchased under the iron rod of experience. It is not till
the want of a green spot is brought painfully home to
people by its loss, that the thought of saving the last
remaining speck of greenery is borne in upon them with
sufficient force to transform the wish into action. For
generations garden after garden has passed into building
land. No one has a right to grudge the wealth or prosperity that has accrued in consequence, but the wish that
the benevolence and foresight of past days had taken a
different bent, and that a more systematic retention of
some of the town gardens had received attention, cannot
be banished.

When Roman civilisation had been swept away in
Britain, and with it all vestiges of the earliest gardens,
there are no vestiges of horticulture until Christianity had
taken hold of the country, and religious houses were
rising up in various parts of the kingdom. The cradle
of modern gardening may be said to have been within
the peaceful walls of these monastic foundations. In no
part of the country were they more numerous than in
and around London, and it is probable that every estab-
lishment had its garden for the supply of vegetables, and
more particularly medicinal herbs. Attached to most
of them, there was also a special garden for the production of flowers for decoration on church festivals. It is
probable that the earliest London gardens were of this
monastic character, and as long as the buildings were
maintained the gardens were in existence. The Grey,
the Black, the White, and the Austin Friars all had
gardens within their enclosures; and the Hospitaller
Orders—the Templars and Knights of St. John—had
large gardens within their precincts. The Temple

Garden is still one of the charms of London, but only the old gateway of the Priory of St. John in Clerkenwell remains, and the garden, with all its historical associations, has long since vanished. It was in a small upper room, " next the garden in the Hospital of St. John of Jerusalem in England, without the bars of West Smythfield," that Henry VII., in the first year of his reign, gave the Great Seal to John Morton, Bishop of Ely, and appointed him Chancellor, and he " carried the seal with him" to his house, Ely Place, hard by.[1] These small references show the picturesque side of such events, the gardens constantly being the background of the scenes.

It is only one more of the regrettable results of the barbarous way in which the Reformation was carried out in England, that the gardens shared the fate of the stately buildings round whose sheltering walls they flourished. It is not easy to picture the desolation of those days: the unkept, uncared-for garden, trodden under foot, makes the forlorn aspect of the despoiled monasteries more pathetic.

London was a city of palaces in Plantagenet times, and the great nobles had their gardens near or surrounding their castles. Bayard's Castle, facing the river for centuries, had its gardens, and there were spacious gardens within the precincts of the Tower when it was the chief royal residence in London, and outside the walls of the City fine dwellings and large gardens were clustered together. Among the most famous in the thirteenth century was the Earl of Lincoln's, purchased from the Dominicans, when they outgrew their demesne in Holborn, and migrated to the riverside, where their

[1] Close Roll, Henry VII.

memory ever lives under their popular name of the Black Friars. Minute accounts of the expenses of this garden are preserved in the Manor Roll, and a very fairly accurate picture of what it was can be pieced together. The chief flowers in it were roses, and the choicest to be found at that date, the sweet-scented double red "rosa gallica," would be in profusion. It might be that, in the shady corners of the garden, periwinkle trailed upon the ground, and violets perfumed the air. White Madonna lilies reared their stately heads among the clove pinks, lavender, and thyme. Peonies, columbines, hollyhocks, honeysuckle, corncockles, and iris, white, purple, and yellow, made no mean show. The orchard could boast of many kinds of pears and apples, cherries and nuts. A piece of water described as "the greater ditch"[1] formed the fish stew where pike were kept and artificially fed. Besides all this, there was a considerable vineyard. It was thought a favourable spot for vines, and the Bishop of Ely's vineyard, the site of which is still remembered by Vine Street, was hard by. A good deal of imagination is now required to conjure up a picture of a vintage in Holborn. Amid the crowd of cabs, carts, carriages, and omnibuses rolling all day over the Viaduct from Oxford Street to the heart of the City, it needs as fertile a brain as that of the poet who pictured the vision of poor Susan as she listens to the song of the bird in Wood Street to call up such a scene. The gardens sloping down to the "bourne" were carefully enclosed—the Earl of Lincoln's by strong wooden palings, that of Ely Place by a thorn hedge with wooden gates fitted with keys and locks.[2] The

[1] MSS. Manor Roll in the Record Office.
[2] MSS. Manor Roll, Archives of Ely Cathedral.

inner gardens, that were specially reserved for the Bishop, the great garden and the "grassyard," were separated by railings and locked doors from the vineyard. The "grassyard" was mown, and a tithe of the proceeds from the sale of the grass paid to the Rector of St. Andrew's, Holborn. The wine produced was more of the character of vinegar, and was also sold; as much as thirty gallons of this "verjuice" was produced in one year. Extra hands were hired to weed and dress the vineyard, and apparently the vineyard entailed a good deal of trouble, and for many years it was let. Think of a warm day in early autumn, clusters of grapes hanging from the twisted vines, men and women in gay colours carrying baskets of ripe fruit to the vats where they were trodden, and the crimson juice squeezed out; the mellow rays of the sinking sun light up the high walls and many towers of the City, and the distant pile of Westminster is half hidden by the mists rising from the river, while there, too, the vintage is in full swing, and the song[1] of the grape-gatherers breaks the stillness of the October evening. Away to the north the landscape is bounded by the wooded heights of Hampstead and Highgate. Most of the country round London then was forest land, and in spite of the changes of centuries a few acres of the original forest remain in Highgate Woods to this day, now owned by the Corporation of London. Between the hills and the city on the north-east lay the marshy ground known as Moorfields, for some 800 years the favourite resort of Londoners wishing to take the air. Gradually this open space has been built over, although a few green patches, such as Finsbury Square, the Artillery Ground, or the more

[1] See Alexander Necham, *De Naturis Rerum*, twelfth century.

distant Bunhill Fields, have remained through the changes time has wrought. This space might have been like one of the other heaths or commons of London, a beautiful open space in the heart of the town, but the supposed exigencies of modern civilisation, with the usual want of foresight, have banished the life-giving fresh air, and the Corporation of London has had to go far afield, to Burnham Beeches and Epping Forest, to supply what once was at its door. Literally at its door, as the busy street of Moorgate recalls the Mayor, Thomas Falconer by name, who in 1415 "caused the wall of the citie to be broken neere unto Coleman Street, and there builded a posterne now called *Moorgate*, upon the Mooreside, where was never gate before. This gate he made for ease of the citizens, that way to passe upon cawseys into the Field for their recreation." [1] The fields in question were at that time a marsh, and though some fifty years later "dikes and bridges" were made, it was many years before the whole moor was drained. The task at one time seemed so difficult that the chronicler Stowe, in 1598, feared that even if the earth was raised until it was level with the city walls it would be "but little dryer," such was the "moorish" nature of the ground. Moorfields was the scene of many curious dramas during its history It was the great place for displays, sham fights, and sports of the citizens. Pepys notes in his Diary, July 26, 1664, that there was much discourse about "the fray yesterday in Moorfields, how the butchers at first did beat the weavers (between whom there hath been ever an old competition for mastery), but at last the weavers rallied and beat them." Such scenes were very frequent, and Moorfields for generations

[1] Stowe, "Survey of London."

was the theatre of such contests.   During the time of the Great Fire, numbers of homeless people camped out there, passing days of discomfort and anxiety about their few remaining household goods.   Pepys in his casual way alludes to them : " 5th September, . . . Into Moorefields (our feet ready to burn, walking through the town among hot coles), and find that full of people, and poor wretches carrying their goods there, and everybody keeping his goods together by themselves (and a great blessing it is to them that it is fair weather for them to keep abroad night and day); drunk there and paid twopence for a plain penny loaf."   The " trained bands " used Moorfields as their exercise ground, and no doubt the prototype of John Gilpin disported himself there. As the fields were drained after 1527 they became more and more the favourite resort of citizens of all ranks. Laid out more as a public garden in 1606, they continued the chief open space of the city until a few generations ago.

The garden of the Drapers' Company was another of the lungs of the City, and the disappearance of the great part of it, also within recent years, is much to be regretted.   This land was purchased by the Company from Henry VIII. after the garden had been made by Thomas Cromwell, Earl of Essex, and forfeited on his attainder.   His method of increasing his garden was simple enough.   He appears to have taken what he wanted from the citizens adjoining, and his all-powerful position at the time left them without redress.   Stowe describes the way this land was filched away.   "This house being finished, and having some reasonable plot of ground left for a garden, hee caused the pales of the gardens adjoining to the north part thereof, on a sudden

to be taken doune, 22 foot to be measured forth right
into the north of every man's ground, a line then to
be drawn, a trench to be cast, a foundation laid, and an
high bricke wall to be builded. My Father had a garden
there, and there was a house standing close to his south
pale; this house they loosed from the ground, and bare
upon Rowlers into my Father's garden 22 foot ere my
Father heard thereof. . . . No man durst goe to argue
the matter, but each man lost his Land."

It is difficult to estimate whether the charitable munificence of the Company is altogether as great a public
benefit, from a health point of view, as retaining some
of the garden for public use would have been. Men are
naturally so conservative, that, because they have been
content to talk and do business, and even search for a
breath of air, in the crowded streets on the hottest
summer days, it has probably never occurred to them
that a few minutes on a seat under shady trees would
have "refreshed their spirits," and the addition of
better air improved their brain powers more effectually.
The idea of a garden city is such a new one that it
is not fair to judge by such standards. Distances are
now much reduced by electricity above and below
ground, so that the necessity of crowding business
houses together to save time is not so all-important.
When the City gardens became built over, no doubt the
newer and more sanitary conditions were felt amply to
compensate for the loss of oxygen given off by the
growing plants, and the preservation of air spaces in the
midst of crowded centres had not occurred to men's
minds.

London four or five hundred years ago must indeed
have needed its gardens. The squalor and dirt of its

cramped streets, the noisy clamour, the rough and un-
couth manners, are unpleasing to realise. The contrast
of the little walled gardens, where the women could sit,
and the busy men find a little quiet from the noise
outside, must indeed have been precious. The pro-
fession of a gardener, however, did not seem to soften
their behaviour, for some of the worst offenders were
gardeners. So serious did the " scurrility, clamour, and
nuisance of the gardeners and their servants," who sold
their fruit and vegetables in the market, become, that
they disturbed the Austin Friars at their prayers in the
church hard by, and caused so much annoyance to the
people living near, that in 1345 a petition, to have these
" gardeners of the earls, barons, bishops, and citizens "
removed to another part of the town, was presented to
the Lord Mayor. Later on, gardening operations in the
City and for six miles round were restricted to freemen
and apprentices of the Gardeners' Company, and the sale
of vegetables was almost exclusively in their hands. Their
guild had power to seize and destroy all bad plants, or
those exposed for sale by unlicensed persons. The
Gardeners' Company, incorporated in 1605, had a second
charter in 1616, and a confirmation of their rights in
1635, and it still remains one of the City companies.

All the smaller householders, even in the crowded
parts, continued to enjoy their little gardens for many
centuries. Even after the spoliation of the monasteries,
the houses rebuilt on their sites had their little enclosures ; and large houses such as Sir William Pawlet's,
on the ground of the Augustine monastery, or later on
Sir Christopher Hatton's on Ely Place, had their gardens
around them. Even now, in the heart of London, a small
row of shabby old houses survives, each with a small garden

attached to it. These are called Nevill Court, from the site having been within the precincts owned by Ralph Nevill, Bishop of Chichester, Chancellor in the time of Henry III., who built a great palace near here. One of the row belongs to the Moravian Mission, or United Brothers, a sect who trace their origin to John Huss. They settled in this house in 1737. This old-world corner opens out of Fetter Lane. A small wooden paling separates the minute strips of blackened garden from a narrow paved pathway. There were many such gardens in this locality less than a century ago. Charles Lamb, when aged six, went to school to a Mr. Bird in Bond Stables, off Fetter Lane, now vanished; and, returning to the spot in 1825, he recalled the early associations: " The school-room stands where it did, looking into a discoloured, dingy garden. . . . Oh, how I remember . . . the truant looks side-long to the garden, which seemed a mockery of our imprisonment." Would that some antiquarian millionaire—if such a combination exists!—might take into his head to preserve Nevill Court, to restore the houses and renovate the gardens, and preserve this relic of Old London, to give future generations some idea of what the smaller dwelling-houses in the old city were like. In most districts these little gardens were the usual appendage to dwelling-houses. Pepys, living in Seething Lane, often mentions his garden. It was there he sat with his wife and taught her maid to sing; it was there he watched the flames spreading over the town at the time of the Great Fire; and in it his money was buried during the scare of the Dutch invasion. So carelessly, indeed, was the money hidden that 100 gold pieces were lost, but eventually most of them recovered by sweeping the grass and sifting the soil. The natural way in which Pepys mentions how other people—Sir W. Batten and Mrs. Turner—during the Fire buried in their city gardens their wine and other goods they could not send to the country, that is, Bethnal Green, only shows how general these little plots were.

Gerard, that delightful old herbalist and gardener to Lord Burghley, in Elizabeth's reign, had his own garden in Holborn. In it flourished no less than some 972 varieties of plants, of which he published a catalogue in 1596. His friend and fellow-botanist, L'Obel, whose name is best remembered by the familiar genus Lobelia, testified that he had seen all the plants on the list actually growing there. The great faith and skill with which these old gardeners attempted to grow in London all the newly-acquired floral treasures, from all parts of the world, is truly touching. To make them "denizons of our London gardens" was Gerard's delight. And this worthy ambition was shared by L'Obel, who looked after Lord Zouche's garden in Hackney; by John Parkinson, author of the delightful work on gardening; and later on, the mantle descended to the Tradescants, who had their museum (the nucleus of the Ashmolean) or "Ark" and garden in Lambeth; by Sir John Sloane, who established the Physic Garden in Chelsea, and numerous others. It is curious to think how many of the plants now familiar everywhere made their first appearance in London. They were not reared elsewhere and brought to the large shows which are arranged in the metropolis to exhibit novelties to the public, but really London-grown. They were foreign importations, little seeds or bulbs, sent home to the merchants trading with the Levant, or brought back by enterprising explorers from the New

World and carefully nurtured in the London gardens, that the citizens "set such store by." There were several of these "worshipful gentlemen" to whom the introduction of flowers is due, and of many a plant Gerard could say with pride, they "are strangers to England, notwithstanding I have them in my garden." Most plants were grown for use, but others "we have them," says Gerard, "in our London gardens rather more for toyes of pleasure than any vertues they are possessed with." Some of the first potatoes introduced were grown in London. Gerard had those in his garden direct from Virginia, and prized them as "a meat for pleasure." Jerusalem artichokes were brought to London by him, and grown there in early days (1617). Parkinson also had them, calling them "Potatos of Canada." Bananas were first seen in England in Johnson's the herbalist's shop in Snow Hill. At a much later date—early in last century—the fuchsia was made known for the first time to Lee, a celebrated gardener, who saw a pot of this attractive plant in the window of a house in Wapping, where a sailor had brought it as a present to his wife. So attached to it was she, that she only parted with it when a sum of eight guineas was offered, besides two of the young rooted cuttings. London can claim so many flowers, it would be tedious to enumerate them all. The first cedars in this country grew in the Chelsea Physic Garden, some of the first orchids at Loddige's Garden in Hackney, and many things have emanated from Veitch's Nursery, or the Botanical Gardens in Regent's Park, or the gardens which used to belong to the Royal Horticultural Society in South Kensington. The chrysanthemum in early days flourished in Stoke Newington, and one of the very first results of cross-fertilisation, which now forms the chief part of scientific garden work, was accomplished by Fairchild, a famous nurseryman at Hoxton, who died in 1730.

This same Thomas Fairchild left a bequest for a sermon, to be preached annually on Whit Tuesday, at St. Leonard's, Shoreditch, on "the Wonderful Works of God in the Creation," which is still delivered, often by most excellent preachers, but to a sadly small and unapprecia-tive congregation. Every opportunity ought to be taken to awaken the interest in these wonders of creation in the vegetable kingdom, and so much might be done in London Parks. They are too frequently merely places of recreation, and until recently but little has been attempted to arouse enthusiasm for the beauties of nature, and to make them instructive as well as attractive. Even in the crowded heart of London a great deal could be effected, and it is a satisfaction to feel that attention is being drawn to the subject and an effort being made in the right direction. In the summer of 1906 a "Country in Town Exhibition" was held in Whitechapel. This novel idea was so successful, and met with such apprecia-tion, that 33,250 people visited the exhibition during the fortnight it was open, besides the hundreds that collected to see H.R.H. Princess Christian perform the opening ceremony. The available space of the White-chapel Art Gallery was filled with plants that would thrive in London; the Office of Works arranged a demonstration of potting; bees at work, aquaria, speci-mens dried by children or drawn in the schools, growing specimens of British plants, such as the dainty bee-orchis, plants and window boxes grown in the district, and such-like, made up the exhibits. Lectures were organised on plant life and nature in London which were largely attended. A series of drawings and plans of the Mile End Road and Shadwell, as they are, and as they might be, were prepared, and the cost of such transformations was worked out. These were exhibited in the hopes of awakening the interest of the Corporation who owns the site of the disused market in Shadwell, and of causing more to be done in the Mile End Road. It appears that with a comparatively small expenditure and ultimate loss, these plans could be realised, and the physical and moral conditions of the whole neighbourhood improved.

Every year it is further to get into the country from the centres of population, and the necessity of improving existing open spaces becomes all the greater. By improving it is not meant to suggest that what are sometimes called improvements should be carried out; grander band-stands, handsome railings, more asphalt paths or stiff concrete ponds. No, it is only more intelligent planting, grouping for artistic effect, and arranging to demonstrate the wonders of nature in spaces already in existence, and to suggest what could be done to cheer and brighten the dark spots of the city.

The country round London has always been a good district for wild flowers; the varied soils, aspects, and levels all go to make it a propitious spot for botanising. Many places now covered with streets were a few generations ago a mass of wild flowers. The older herbalists— Gerard, Johnson, and their friends—used to search the neighbourhood of London for floral treasures, and incidentally in their works the names of these friends, such as Mr. James Clarke and Mr. Thomas Smith, "Apothecaries of London," and their "search for rare plants" are mentioned. Gerard was constantly on the watch,

and records plants seen in the quaintest places, such as
the water-radish, which he says grew "in the joints or
chincks amongst mortar of a stone wall that bordereth
upon the river Thames by the Savoy in London, which
yee cannot finde but when the tide is much spent."
Pennyroyal "was found on the common near London
called Miles ende," "from whence poore women bring
plentie to sell in London markets." The rare adders-
tongue and great wild valerian grew in damp meadows,
the fields abounded with all the more common wild
flowers, and such choice things as the pretty little
"ladies' tresses," grew on the common near Stepney,
while butcher's broom, cow wheat, golden rod, butterfly
orchis, lilies of the valley and royal fern, wortleberries and
bilberries covered the heaths and woods of Hampstead and
Highgate. Many another flower is recorded by Gerard,
who must have had a keen and observant eye which
could spot a rare water-plant in a ditch while attending
an execution at Tyburn! yet he meekly excuses his want
of knowledge of where a particular hawkweed grew,
saying, "I meane, God willing, better to observe heerafter,
as oportunitie shall serve me." That power of observa-
tion is a gift to be fostered and encouraged, and were
that achieved by education in Council Schools, a great
success would have been scored, and probably it would
be more fruitful in the child's after life than the scattered
crumbs from countless subjects with which the brain is
bewildered. The wild flowers could still be enticed
within the County of London, and species, which used
to make their homes within its area, might be induced
at least to visit some corners of its parks. The more
dingy the homes of children are, the more necessary it
must be to bring what is simple, pure, and elevating to

their minds, and modern systems of teaching are realising this. If public gardens can be brought to lend their aid in the actual training, as well as being a playground, they will serve a twofold purpose. An old writer quaintly puts this influence of plant life. "Flowers through their beautie, varietie of colour and exquisite forme, do bring to a liberall and gentle manly mind, the remembrance of honestie, comelinesse, and all kindes of vertues. For it would be an unseemly and filthie thing, as a certain wise man saith, for him that doth looke upon and handle faire and beautifull things, and who frequenteth and is conversant in faire and beautifull places, to have his mind not faire but filthie and deformed."

It is not possible for all London children to get into the country now that it is further away, so the more of nature, as well as true artistic gardening, they can be shown in the parks the better. It used in olden days to be the custom, among other May Day revels, to go out to the country round London and enjoy the early spring as the Arabs do at the present time, when they have the fête of "Shem-en-Nazim," or "Smelling the Spring." "On May day in the morning, every man, except impediment, would walk into the Sweet Meddowes and green woods, there to rejoyce their spirits with the beauty and Savour of sweet Flowers, and with the harmonie of Birdes, praising God in their kinde."[1] It would surprise many people to learn how many birds still sing their praises within the parks of London, although the meadows and other delights have vanished. This serves to encourage the optimist in believing in the future possibilities of London Parks.

There is no "park system" in England as in the

---

[1] Stowe's "Survey of London."

United States of America, where each town provides, in addition to its regular lines of streets, and its main thoroughfares leading straight from the centre to the more suburban parts, a complete system of parks. The more old-fashioned town of Boston was behind the rest, although it contained a few charming public gardens in the heart of the town. Of late years large tracts of low-lying waste grounds have been filled up, and one piece connected with another, until it, too, rejoices in a complete "park system." Chicago, Pittsburgh, and all these modern towns of rapid growth possess a well-ordered "park system." The conditions, the natural aspect of the country, and the climate are so unlike our own that no comparison is fair. Like everything else in the United States, they are on a large scale, and while there is much to admire, and something to learn, there is very little in the points in which they differ from us that could be imitated. London parks and open spaces, taken as a whole, are unrivalled. The history and associations which cluster round each and all of them, would fill volumes if recorded facts were adhered to; and if the imagination were allowed to run riot within the range of possibility, there would be no limit. Things which have grown gradually as circumstances changed can have no system. Their variety and irregularity is their charm, and no description of either the parks, gardens, or open spaces of London can be given as a whole. Each has its own associations, its own history, and to glance at some of London's bright spots and tell their stories will be the endeavour of these pages.

CHAPTER II

HYDE PARK

The Park shone brighter than the skyes,
Sing tan-tara-rara-tantivee,
With jewels and gold, and Ladies' eyes,
That sparkled and cry'd come see me :
Of all parts of England, Hide Park hath the name,
For coaches and Horses and Persons of fame,
It looked at first sight, like a field full of flame,
Which made me ride up tan-tivee.
—News from Hide Park, an old ballad, *c.* 1670.

IN writing about London Parks the obvious starting-point seems to be the group comprising Hyde, Green, and St. James's Parks, which are so intimately connected with London life to-day, and have a past teeming with interest. What changes some of those elms have witnessed! Generation after generation of the world of fashion have passed beneath their shades. Dainty ladies with powder and patches have smiled at their beaux, perhaps concealing aching hearts by a light and careless gaiety. Stately coaches and prancing horsemen have passed along. Crowds of enthusiasts for various causes have aired their grievances on the green turf. Brilliant reviews and endless parades have taken place on the wide open spaces; games and races have amused thousands

of spectators.   In still earlier times there was many a
day's good sport after the deer, or many a busy hour's
ploughing the abbey lands of the then Manor of Hyde.
Scene after scene can be pictured down to the present
time, when, after centuries of change, the enjoyment of
these Parks remains perhaps one of the most treasured
privileges of the Londoner.

In tracing the history of their various phases, the
survival of many features is as remarkable as the disappearance of others.   The present limits on the north
and east, Bayswater Road and Park Lane, have suffered
no substantial alteration since the roads were known as
the Via Trimobantina and the Watling Street in Roman
times.   The Watling Street divided, and one section
followed the course of the present Oxford Street to the
City ; the other, passing down the line of Park Lane,
crossed St. James's Park, and so to the ford over the
Thames at Westminster.   The Park was never common
or waste land, but must have been cleared and cultivated
in very early times.   In Domesday Survey the Manor
was in plough and pasture land, with various "villains"
and peasants living on it.   The Thames was the southern
boundary of the Manor of "Eia," which was divided
into three parts, one being Hyde, the site of the existing
Hyde Park, the other two Ebury and Neate.   Although now forgotten, the latter name was familiar for
many centuries.   When owned by the Abbots of Westminster, the Manor House by the riverside was of some
importance, and John of Gaunt stayed there.   Famous
nurseries and a tea garden, "the Neate houses," marked
the spot in the eighteenth century.

Until the stormy days of the Reformation these lands
remained much the same.   Owned by the Abbey of

Westminster, they were probably well cultivated by their tenants, and doubtless the game with which they abounded from early times afforded the Abbot some pleasant days' sport and tasty meals. The first time any of the Manor became part of the royal demesne, was when the Abbot Islip exchanged 100 acres of what is now St. James's Park, adjoining the royal lands, for Poughley in Berkshire, with Henry VIII. in 1531-2. This Abbot, who had an ingenious device to represent his name—a human eye and a cutting or "slip" of a tree —died in the Manor House of Neate or Neyte the same year. He gave up the lands from Charing Cross "unto the Hospital of St. James in the fields" (now St. James's Palace), and the meadows between the Hospital and Westminster. Five years later, when the upheaval of the dissolution of the monasteries was taking place, the monks of Westminster were forced to take the lands of the Priory of Hurley—one of their own cells just dissolved—in exchange for the rest of the manor. Henry VIII., who loved sport, found these lands first-rate hunting-ground. From his palace at Westminster, through Hyde Park, right away to Hampstead, he had an almost uninterrupted stretch of country, where hares and herons, pheasants and partridges, could be pursued and preserved "for his own disport and pastime." Hyde Park was enclosed, or "substancially empayled," as an old writer states, and a large herd of deer kept there, and various proclamations show that the right of sport had to be jealously guarded.

What a gay scene must Hyde Park have often witnessed in Elizabeth's reign. The Queen, when not actually joining in the chase, watched the proceedings from the hunting pavilion, or "princelye standes therein,"

and feasted the guests in the banqueting-house.   There
were brilliantly caparisoned horses, men and women in
costly velvets and brocades, stiff frills, plumed hats and
embroidered gloves.   Picture the *cortège* entering by the
old lodge, where now is Hyde Park Corner, the honoured
guest, for whom the day's sport was inaugurated—such
as John Casimir, son of the Elector Palatine, who showed
his skill by killing a particular deer out of a herd of 300
—surrounded by some of his foreign attendants, and
escorted by all the court gallants of the day.

The Park must then have been as wild as the New
or Sherwood Forests of to-day.   The tall trees, with
their sturdy stems, were then untouched by smoky
air, the sylvan glades and pasture lands had no distant
vistas of houses and chimneys to spoil their rural aspect,
while far off the pile of the buildings of Westminster
Abbey—without the conspicuous towers, which were
not finished till 1714—might be seen rising beyond the
swamps and fens of St. James's Park.   Hyde Park on
a May evening even now is still beautiful, if looked at
from the eastern side across a golden mist, against
which the dark trees stand up mysteriously, when a
glow of sunset light seems to transform even ragged
little Cockney children into fairies.   It wants but little
imagination to see that same golden haze peopled with
hunstmen, and to hear the sound of the horn instead
of the roar of carriages.

The next scene which can be brought vividly before
the mind's eye is very different from the last pageant.
These are troublous times.   The monarch and his
courtiers are occupied in far other pursuits than hunting deer.   Charles I. was fighting in the vain endeavour
to keep his throne, and Londoners were preparing to

defend the city. Hyde Park and Green Park became
the theatre of warlike operations. Forts were raised
and trenches were dug. Two small forts, one on Constitution Hill and one near the present Mount Street
in Hyde Park, were made, but the more important were
those on the present sites of the Marble Arch and of
Hamilton Place. The energy displayed on the occasion
is described by Butler in "Hudibras," and the part taken
by women in the work. Like the "sans culottes" of
the French Revolution, they helped with their own
hands.

> " Women, who were our first apostles,
> Without whose aid w' had all been lost else ;
>
> .   .   .   .
>
> March'd rank and file, with drum and ensign,
> T' entrench the city for defence in ;
> Rais'd rampires with their own soft hands,
> To put the enemy to stands ;
> From ladies down to oyster-wenches
> Labour'd like pioneers in trenches,
> Fell to their pickaxes and tools,
> And helped the men to dig like moles."
>   —BUTLER'S " *Hudibras.*"

The picture of their sombre garments, neat-fitting
caps, and severe faces, the close-cropped hair and stern
looks of the men, working with business-like determination, stands out a striking contrast to the gay colours
and cheerful looks of the company engaged in the chase.

The darker trees and sheltered corners of Hyde Park
afforded covert for the wary "Roundhead" to lie in
ambush for the imprudent Loyalist carrying letters to the
King. On more than one occasion the success was on his
side, and the bearer of news to his royal master was way-
laid, and the papers secured. The culminating scene of

this period must have been when Fairfax and the Parliamentary army marched through Hyde Park in 1647, and were met by the solemn procession of the Mayor and Sheriffs of the City of London.

Dismal days for the Parks followed. Although the Parks had been declared the property of the Commonwealth, it was from no wish to use them for sport or recreation. During the latter years of Charles the First's reign Hyde Park had become somewhat of a fashionable resort. People came to enjoy the air and meet their friends, and it was less exclusively reserved for hunting. Races took place, both foot and horse ; crowds collected to witness them, and ladies, with their attendant cavaliers, drove there in coaches, and refreshed themselves at the "Cake House" with syllabubs. This latter was the favourite drink, made of milk or cream whipped up with sugar and wine or cider. But the Puritan spirit, which was rapidly asserting itself, soon interfered with such harmless amusements. In 1645 the Parks were ordered to be shut on the Lord's Day, also on fast and thanksgiving days. In 1649 the Parks, together with Windsor, Hampton Court, Greenwich, and Richmond, were declared to be the property of the Commonwealth, and thrown open to the public. But this did not lead to greater public enjoyment of Hyde Park. Far from it, for only three years later it was put up to auction in three lots. The first lot was the part bounded on one side by the present Bayswater Road, and is described as well wooded ; the second, the Kensington side, was chiefly pasture ; the third, another well-wooded division, included the lodge and banqueting-house and the Ring where the races took place. This part was valued at more than double the two others, and was purchased by

Anthony Dean, a ship-builder, for £9020, 8s. 2d. This business-like gentleman presumably reserved the use of the timber for his ships, and let out the pasture. His tenant proceeded to make as much as he could, and levied a toll on all carriages coming into the Park. On some occasions he extorted 2s. 6d. from each coach. In 1653 John Evelyn in his diary complains on April 11 that he "went to take the aire in Hide Park, when every coach was made to pay a shilling, and every horse six-pence, by the sordid fellow who had purchas'd it of the State, as they were call'd." Cromwell himself was fond of riding in the Park, and crowds thronged him as he galloped round the Ring. More than one plot was made against the life of Cromwell, and the Park was considered a likely place in which to succeed. On one occasion the would-be assassin joined the crowd, which pursued the Protector during his ride, ready, if at any moment he galloped beyond the people, to dash at him with a fatal blow. The plotter had carefully filed the Park gate off its hinges so as to make good his own escape. It is a curious fact that Cromwell more nearly met his death in Hyde Park by accident than by design. He was pre-sented with some fine grey Friesland horses, by the Duke of Holstein, and insisted on driving the spirited animals himself. They bolted, he was thrown from the box, and his pistol went off in his pocket, "though without any hurt to himself"!

The Ring, where all these performances took place, was situated to the north-east of where the Humane Society's house, built in 1834, now stands, near the Serpentine. There are a few remains of very large elm trees still to be seen, which probably shaded some of the company assembled to watch the coaches driving round and round

the Ring, or cheer the winner of a hotly-contested race.
Even during the sombre days of the Commonwealth sports
took place in the Park, but with the Restoration it became
much more the resort of all the fashionable world and the
scene of many more amusements. The parks were still
in those days for the Court and the wealthy or well-to-do
citizens only. Probably to many of the rabble and poorer
Londoners the nearest view obtained of Hyde Park would
be the tall trees within its fence or wall, which formed a
background to the revolting but most engrossing of
popular sights, the horrors of the gallows at Tyburn.
The idea of giving parks as recreation grounds for the
poor is such a novel one that no old writer would think
of noticing their absence in an age when bull-baiting and
cock fights were their highest form of amusement.

The Ring was an enclosure with a railing round it
and a wide road. It is described as "a ring railed in,
round wch a gravel way, yt would admitt of twelve if
not more rowes of Coaches, wch the Gentry to take the
aire and see each other Comes and drives round and
round; one row going Contrary to each other affords a
pleaseing diversion."

The gay companies who assembled to drive round
and round the Ring, or watch races, sometimes met with
unusual excitement. On one occasion Hind, a famous
highwayman, for a wager rode into the Ring and robbed
a coach of a bag of money. He was hotly pursued across
the Park, but made his escape, "riding by St. James's,"
which then, and until a much later date, was a sanctuary,
and no one except a traitor could be arrested within it.
So narrow an escape from justice did he have that he is
said to have exclaimed, "I never earned £100 so dear in
all my life!"

Numberless entries in Pepys' Diary describe visits to Hyde Park. His drives there in fine and wet weather, the company he met, whether his wife looked well or was in a good or ill temper, and the latest gossip the outing afforded, are all noted. Many times he regrets not having a coach of his own, and does not conceal the feelings of wounded pride it occasioned. Once he naïvely explains that having taken his wife and a friend to the Park "in a hackney," and they not in smart clothes, he "was ashamed to go into the tour [Ring], but went round the Park, and so, with pleasure, home." His delight when he possessed a coach is unbounded. He made frequent visits to the coach-builder, and watched the final coat of varnish to "make it more and more yellow," and at last on May Day, 1669, he describes his first appearance in his own carriage : "At noon home to dinner, and there find my wife extraordinary fine with her flowered tabby gown that she made two years ago, now laced exceeding pretty, and indeed was fine all over, and mighty earnest to go ; though the day was very lowering ; and she would have me put on my fine suit, which I did. And so anon, we went alone through the town with our new liveries of serge and the horses' manes and tails tied with red ribbons, and the standards gilt with varnish, and all clean, and green reines, that people did mightily look upon us ; and the truth is I did not see any coach more pretty, though more gay than ours, all that day . . . the day being unpleasing though the Park full of Coaches, but dusty, and windy, and cold, and now and then a little dribbling of rain ; and what made it worse, there were so many hackney coaches as spoiled the sight of the gentlemen's, and so we had little pleasure. But here was Mr. Batelier and his sister in a borrowed coach by

themselves, and I took them and we to the lodge : and at
the door did give them a syllabub and other things, cost
me 12s. and pretty merry."

What an amusing picture, not only of Hyde Park in
1669 but of human nature of all time !—the start, the
pride and delight with their new acquisition, the little
annoyances, the marred pleasures, the ungenerous dislike
of the less fortunate who could not afford coaches of
their own, whose ranks he had swelled the very last drive
he had taken.   Then the little kindness and the refreshment, so that the story ends merrily.

The " Lodge " is but another name for the " Cheese-
cake House " or " Cake House," or as it was sometimes
called from the proprietor, the Gunter of those days,
" Price's Lodge."   This house, which was a picturesque
feature, stood near the Ring, on the site of the present
building of the Humane Society, and must have been
the scene of many amusing incidents in the lives of
those who graced the Ring, in the seventeenth and
eighteenth centuries.   A little stream ran in front of
it, and the door was approached over planks.   White
with beams of timber, latticed windows, and gabled roof,
a few flowers clustering near, with the water flowing by
its walls, the old house gave a special charm and rural
flavour to the tarts and cheesecakes and syllabub with
which the company regaled themselves.

The gay sights and sounds in Hyde Park were
silenced during those terrible weeks, when the Great
Plague spread death and destruction through London.
As the summer advanced, and the havoc became more
and more appalling, many of the soldiers quartered in the
city, were marched out to encamp in Hyde Park.   At
first it seemed as if they would escape the deadly

scourge, but the men were not accustomed to the rough quarters, and soon succumbed.

> " Our men (ere long) began to droop and quail,
> Our lodgings cold, and some not us'd thereto,
> Fell sick, and dy'd, and made us more adoe.
> At length the Plague amongst us 'gan to spread,
> When ev'ry morning some were found stark dead ;
> Down to another field the sick we t'ane,
> But few went down that e'er came up again."

Thus all through the autumn of that terrible year the Park was one of the fields of battle against the relentless foe. The contemporary poet, whose lines have been quoted, describes the return of the few saddened survivors to the " doleful " city. They had lingered through the cold and wet until December, and surely the Park has no passage in its history more piteous and depressing than the advent of those frightened men who came with " heavy hearts," " fearing the Almighty's arrows," only to be overtaken by the terror in their plague-stricken camp.

Hyde Park has witnessed other gloomy pictures from time to time. Although the colouring of fashion and romance has endeavoured to make these incidents less repulsive, duels cannot be otherwise than distressing to the modern sense. For generations Hyde Park was a favourite place in which to settle affairs of honour. The usual spot is described by Fielding in " Amelia." The combatants walked up Constitution Hill and into Hyde Park " to that place which may properly be called the Field of Blood, being that part a little to the left of the Ring, which Heroes have chosen for the scene of their exit out of this World." One of the most famous duels was that fought between Lord Mohun and the Duke of

c

Hamilton on November 15, 1712, which resulted in the death of both the combatants—the Duke, whose loss was a great blow to the Jacobite cause in Scotland, and the Whig opponent. All through the eighteenth century Hyde Park was frequently the place in which disputes were settled, and one of the last duels recorded, which resulted in the death of Captain Macnamara (his antagonist, Colonel Montgomery, being tried for manslaughter, but acquitted), although fought on Primrose Hill, originated in Hyde Park. The cause of quarrel was that the dogs of these two gentlemen fought while out with them in the Park, whereupon the respective masters used such abusive language to each other that the affair had to be settled by a duel.

Military displays, for which Hyde Park is still famous, have taken place there from early times. The works of defence were thrown up, and Fairfax and the Parliamentary army arrived there in the times of civil strife, but soon after the Restoration Charles II. had a peaceful demonstration, and there reviewed his Life Guards. Again, in September 1668, there was a more brilliant review, when the Duke of Monmouth took command of the Life Guards, and the King and Duke of York were both present. Pepys was there, and wrote, "It was mighty noble, and their firing mighty fine, and the Duke of Monmouth in mighty rich clothes; but the well ordering of the men I understand not."

When, in 1715, the fear of a general Jacobite rising induced the Whigs to take serious precautions, Hyde Park became a camp from July till November. During a similar scare in 1722 troops were again quartered there, and the camp became the centre of popular attraction; gaiety and frivolity were the order of the day, rather than business or watchfulness. The Park was also used as a camp for six regiments of militia at the time of the Gordon Riots in 1780. All through George III.'s long reign reviews were frequent, and one of the most popular was that held by the Prince Regent before the allied sovereigns, the Emperor of Russia and King of Prussia, in June 1814. Blücher was the popular hero on the occasion, and when he afterwards appeared in the Park he was so mobbed by the crowd, enthusiastic to see something of "Forwärts," as he was familiarly named, that he had to defend himself against their rough treatment.

When the Park was again in the King's hands after the Restoration, a Keeper was once more appointed, who was responsible for its maintenance. From the time of Henry VIII. various well-known people had filled the office of Keeper. The first in Henry VIII.'s time was George Roper, succeeded in 1553 by Francis Nevill, and in 1574 by Henry Carey, first Lord Hunsdon, while in 1607 Robert Cecil, Earl of Salisbury, was appointed, and Sir Walter Cope held the office conjointly with him from 1610. The name of the first Keeper after the Restoration, James Hamilton, is well remembered by the site of his house and ground, which are still known as Hamilton Place and Gardens. He was allowed to enclose 55 acres of park, and to use it as an orchard on the condition that he sent a certain quantity of the cider produced from it to the King. In his time a brick wall was built round the Park, and it was re-stocked with deer. The wall was rebuilt in 1726, and not replaced by railings until a hundred years later. These iron railings were pulled down by the mob in 1866, after which the present ones were set up. The deer, which formerly ranged all over the Park, were in course of time confined to a small area on the north-west side, called Buckdean Hill. They were kept for sport during the first half of the eighteenth century, and the last time royalty took part in killing deer in the Park was probably in 1768. The exact date of the disappearance of all the deer is difficult to ascertain. They are remembered by some who saw them towards the end of the thirties, but by 1840 or soon after they were done away with.

The roads in Hyde Park must have been rather like South African tracks at the present day, and driving at night was not free from danger even at a comparatively late date. Attacks from highwaymen were to be feared. Horace Walpole was robbed in November 1749, and the pistol shot was near enough to stun though not otherwise to injure him. The Duke of Grafton had his collar bone broken, and his coachman his leg, some ten years earlier, when, on his way from Kensington to "the New Gate to make some visits towards Grosvenor Square, the Chariot through the darkness of the Night was overset in driving along the Road and " fell " into a large deep pit."

Soon after William III. purchased Kensington Palace from the Earl of Nottingham in 1691, he commenced making a new road through the Park. This became known as the King's Road, or "Route du Roi": a corruption of the latter is Rotten Row, the name now given to King William's Drive. In the eighteenth century it was called the King's Old Road, and the one which George II. made to the south of it was called the King's New Road. When this was finished in 1737, it was intended to turf the older "Rotten Row," but this plan was never carried out. The old road was much thought of at the time it was made, and the lighting of it up at night with 300 lamps caused wonder to all beholders.

A young lady, Celia Fiennes, describes the road in her diary about 1695. "Y^e whole length of this parke there is a high Causey of a good breadth, 3 Coaches may pass, and on each side are Rowes of posts on wch are glasses—Cases for Lamps wch are Lighted in ye Evening and appeares very fine as well as safe for ye passenger. This is only a private roade ye King had wch reaches to Kensington, where for aire our Great King W^{m.} bought a house and filled it for a Retirement wth pretty gardens."

The road was in bad repair before the new one was in good order, and Lord Hervey, writing in 1736, says it had grown "so infamously bad" as to form "a great impassable gulf of mud" between London and Kensington Palace. "There are two ways through the Park, but the new one is so convex, and the old one is so concave, that by this extreme of faults they agree in the common of being, like the high road, impassable."

One of the most striking features of Hyde Park to-day is the long sheet of water known as the "Serpentine," but this was a comparatively late addition to the attractions of the Park. From earliest times there was water. The deer came down to drink at pools supplied by fresh springs. The stream of the West Bourne flowed across the Park from north to south, leaving it near the present Albert Gate. Near there it was spanned by a bridge, from which the hamlet of Knightsbridge derived its name. The water in the Park was used to supply the West End of London as houses began to be built further from the City, and Chelsea was also supplied from it. The Dean and Chapter of Westminster

had a right to the use of the water from the springs in
the Park, and the history of their privilege is recorded
on a stone which stands above "the Dell" on the north-
east of the bridge across the end of the Serpentine.   The
inscription states that a supply of water by a conduit was
granted to the Abbey of Westminster by Edward the
Confessor, and the further history of the lands, which
passed into Henry VIII.'s hands at a time when all church
property was in peril of seizure, is neatly glossed over as
the " manor was resumed by the Crown in 1536."   The
use of the springs, however, was retained by the Abbey,
and confirmed to them by a charter of Elizabeth in 1560.
Later on the privilege was withdrawn, and in 1663 the
Chelsea Waterworks were granted the use of all the
streams and springs of Hyde Park.   They made in
1725 a reservoir on the east side of the Park, opposite
Mount Street.   The sunk garden, with the Dolphin
Fountain, the statue in Carrara marble, and the basin of
Sicilian marble, by A. Munro, was made in 1861 on the
site of this reservoir, which was abandoned two years
earlier.   It has been stated that this sunk garden was
a remnant of the forts of Cromwell's time, one small
one having been near here, but the history of the Chelsea
Waterworks reservoir must have been unknown to those
who believed the tradition.   It contained a million and a
half gallons of water, and was protected by a wall and
railings, as suicides were once said to have been frequent.
When the Serpentine was made by Queen Caroline, considerable compensation had to be paid to the Waterworks
Company.

In this age of experiments in plant growing, when
American writers glow with enthusiasm on the wonders
of the "New Earth," and when science has transformed

DOLPHIN FOUNTAIN, HYDE PARK

the dullest operations of farming and gardening into
fields for enterprise and treasuries of possible discoveries,
it is humiliating to find the water in Hyde Park being
used for like experiments as long ago as 1691–92.
Stephen Switzer, a gardener, who would have been described by his contemporaries as a " lover of ingenuities,"

DOLPHIN FOUNTAIN IN HYDE PARK

was fond of indulging in speculations, and studied the
effect of water on plants. He quotes a series of experiments made by Dr. Woodward on growing plants
entirely in water, or with certain mixtures. For fifty-two
days during the summer of 1692 he carefully watched
some plants of spearmint, which were all "the most
kindly, fresh, sprightly Shoots I could chuse," and were
set in water previously weighed. For this trial he selected
" Hyde Park Conduit water "—one pure, another had an

ounce and a half of common garden earth added to it, a
third was given an equal quantity of garden mould, and
a fourth was kept on " Hyde Park water distilled." The
results in growth, and the quantity of water absorbed,
were carefully noted at the end of the time.

When Queen Caroline conceived the idea of throw-
ing the ponds in Hyde Park into one, and making a
sheet of water, the school of " natural " or " landscape "
gardening was becoming the rage. Bridgeman, a well-
known garden designer, who had charge of the royal
gardens, has the credit of having invented the " ha-ha "
or sunk fence, and thus led the way for merging gardens
into parks. Kent, who followed him, went still further.
He, Horace Walpole said, " leaped the fence, and saw
that all Nature was a garden." The fashions in garden
design soon change, and the work of a former generation
is quickly obliterated. William III. brought with him
the fashion of Dutch gardening, and laid out Kensington
Gardens in that style. Switzer, writing twenty-five years
later, says the fault of the Dutch gardeners was " the
Pleasure Gardens being stuffed too thick with Box";
they " used it to a fault, especially in England, where
we abound in so much good Grass and Gravel." London
and Wise, very famous nursery gardeners, who made
considerable changes at Hampton Court, and laid out
the grounds of half the country seats in England, had
charge of Kensington Palace Gardens, and housed the
" tender greens " during the winter in their nurseries
hard by. These celebrated Brompton nurseries were so
vast that the Kensington plants took up " but little
room in comparison with " those belonging to the firm.
Queen Mary took great interest in the new gardens.
" This active Princess lost no time, but was either

measuring, directing, or ordering her Buildings, but in Gard'ning, especially Exoticks, she was particularly skill'd, and allowed Dr. Pluknet £200 per ann. for his Assistance therein." After his queen's death William III. did no more to the gardens, but they were completed by Queen Anne. She appointed Wise to the chief care of the gardens, and when in 1712 rules for the "better keeping Hyde Park in good Order" were drawn up, and people were forbidden to leap the fences or ditches, or to ride over the grass, a special exception was made in favour of Henry Wise. Switzer, in tracing the history of gardening to his day (1715), praises the "late pious Queen, whose love to Gardening was not a little," for "Rooting up the *Box*, and giving an *English* Model to the old-made Gardens at *Kensington*; and in 1704 made that new garden behind the Green-House, that is esteemed amongst the most valuable Pieces of Work that has been done any where. . . . The place where that beautiful Hollow now is, was a large irregular Gravel-pit, which, according to several Designs given in, was to have been filled, but that Mr. Wise prevailed, and has given it that surprizing Model it now appears in. As great a Piece of Work as that whole Ground is, 'twas near all completed in one Season, (viz.) between Michaelmas and Lady Day, which demonstrates to what a pitch Gard'ning is arrived within these twenty or thirty years."

When William III. purchased Kensington Palace, the grounds covered less than thirty acres. Under the management of Wise, in Queen Anne's time, more was added, and the Orangery was built in 1705. Few people know the charms of this old building, which stands to the north of the original garden, and which

future alterations may once more bring more into
sight.   As the taste for gardening changed from the
shut-in gardens of the Dutch style to the more extended places of Wise, the garden grew in size.   Again,
when Bridgeman was gardener, Queen Caroline, wife of
George II., wished to emulate the splendour of Ver-
sailles, and 300 acres were taken from Hyde Park to
add to the Palace Garden.   Bridgeman made the sunk
fence which is still the division between Kensington
Gardens and the Park; and with the earth which was
taken out a mount was made, on which a summer-house
was erected.   This stood nearly opposite the present end
of Rotten Row, and though it has long since ceased to
exist, the gate into the Gardens is still known as the
Mount Gate.   Kent, who succeeded Bridgeman, continued the planting of the avenues and laying out of the
Gardens, and the greater part of his work still remains.
The Gardens were reduced in size when the road was
made from Kensington to Bayswater, and the houses
along it built about seventy years ago, and the exact
size is now 274 acres.   Queen Caroline would have
liked to take still more of the Parks for her private use;
but when she hinted as much to Walpole, and asked the
cost, he voiced public opinion when he replied, "Three
crowns."

The fashion of making sheets of artificial water
with curves and twists, instead of a straight, canal-like
shape, was just taking the public fancy, when Queen
Caroline began the work of converting the rather marshy
ponds in Hyde Park into a "Serpentine River."   The
ponds were of considerable size, and in James I.'s time
there were as many as eleven large and small.   Celia
Fiennes, the young lady who kept a diary in the time of

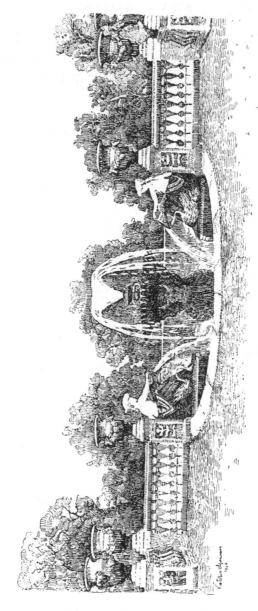

FOUNTAINS AT THE END OF THE SERPENTINE

William and Mary, which has been already quoted, after describing the Ring, says, " The rest of the park is green, and full of deer ; there are large ponds with fish and fowle." The work of draining the ponds and forming a river was begun in October 1730, under the direction of Charles Withers, Surveyor-General of the Woods and Forests. The cost of the large undertaking was supposed to come out of the Queen's privy purse, and it was not until after her death that it was found that Walpole had supplemented it out of the public funds. The West Bourne supplied the new river with sufficient water for some hundred years, after which new arrangements had to be made, as the stream had become too foul. The water supply now comes from two sources—one a well 400 feet deep at the west end of the Serpentine, where the formal fountains and basins were made, about 1861, in front of the building of Italian design covering the well. The sculptured vases and balustrade with sea-horses are by John Thomas. The water in the well stands 172 feet below the ground level, and the depth is continually increasing. It is pumped up to the " Round Pond," and descends by gravity. The second supply comes from a well 28 feet deep in the gravel on " Duck Island," in St. James's Park. The water, which is 19 feet below the surface, remains constant, that level being the same as the water-bearing stratum of the Thames valley in London. It is pumped up to the Serpentine, and returns to the lake in St. James's Park, supplying the lake in the gardens of Buckingham Palace on the way. The deep well provides about 120,000 gallons, and the shallow about 100,000 a day. The " Round Pond "—which, by the way, is not round—affords the greatest delight to the owners, of all ages, of miniature yachts of all sizes. There are the large boats with skilful masters, which sail triumphantly across the placid waters, and there are the small craft that spend days on the weeds, or founder amid "waves that run inches high," like the good steamship *Puffin* in Anstey's amusing poem. When the weeds are cut twice every summer, many pathetic little wrecks are raised to the surface, perchance to be restored to the expectant owners.

Skating was an amusement in Hyde Park even before the Serpentine existed, and the older ponds often presented a gay scene in winter, although it was on the canal in St. James's Park that the use of the modern skate is first recorded in Charles II.'s time.

During the last hundred years Hyde Park has frequently been disturbed by mobs and rioters, until it has become the recognised place in which to air popular discontent in any form, or to ventilate any grievance. The first serious riot took place at the funeral of Queen Caroline, in 1821. To avoid any popular demonstration of feeling, it was arranged that the funeral procession should not pass through the City. The Queen had died at Brandenburgh House, and was to be interred at Brunswick. Instead of going straight by way of Knightsbridge and Piccadilly, a circuitous route by Kensington, Bayswater, Islington, and Mile End was planned. On reaching Kensington Church, the mob prevented the turn towards Bayswater being taken. Hyde Park was thronged with an excited crowd, trying to force the escort to go the way it wished. At Cumberland Gate quite a severe encounter took place, in which the Life Guards twice charged the mob. Further down Oxford Street were barricades, and to avoid further rioting the procession eventually had to take the people's route, passing quietly down to the Strand and through the City.

The occasion of the Reform Bill riot in 1831, when the windows were smashed in Apsley House, is well known, and from 1855 to 1866 Hyde Park witnessed many turbulent demonstrations. The first occasion was in July 1855 against Lord Robert Grosvenor's " Sunday Trading Bill," when some 150,000 people assembled, and various scenes of disturbance took place. More or less serious riots were of frequent occurrence, until they culminated in the Reform League riot in July 1866, when the railings between Marble Arch and Grosvenor Gate " were entirely demolished, and the flower-beds were ruined." The flower-beds had not been long in existence when they were wantonly damaged by the mob.

The idea of introducing flowers into the Park began about 1860, and the long rows of beds between Stanhope Gate and Marble Arch were made about that time, when Mr. Cowper Temple was First Commissioner of Works. They were made when " bedding out " was at the height of its fashion, when the one idea was to have large, glaring patches of bright flowers as dazzling as possible, or minute and intricate patterns carried out in carpet bedding. Now this plan has been considerably modified. The process of alteration has been slow, and the differences in some cases subtle, but the old stiffness and crudeness has been banished for ever. The harmony of colours, and variety of plants used, are the principal features in the present bedding out. It seems right that the Royal Parks should lead the way in originality and beauty, and undoubted success is frequently achieved, although even the style of to-day has its opponents. The chief objection from the more practical gardeners is the

AUTUMN BEDS, HYDE PARK

putting out of comparatively tender plants in the summer months, when the same general effect could be got with a less expenditure both of money and plants. But on the other hand numbers of people come to study the beds, note the combinations, and examine the use of certain plants which they would not otherwise have the opportunity of testing. The public who enjoy the results, and often those who most severely criticise, do not know the system on which the gardening is carried out. Many are even ignorant enough to suppose that the whole bedding out is contracted for, and few know the hidden recesses of Hyde Park, which produces everything for all the display, both there and in St. James's Park. The old place in which all necessary plants were raised was a series of greenhouses and frames in front of Kensington Palace. The erection of these pits and glass houses completely destroyed the design of the old garden, although even now the slope reveals the lines of the old terraces; and they entirely obscure the beauty of the Orangery. A few years ago three acres in the centre of Hyde Park were taken, on which to form fresh nurseries. Gradually better ranges have been built, and soon the old unsightly frames at Kensington will disappear. The new garden is so completely hidden that few have discovered its whereabouts. The ground selected lies to the north-west of the Ranger's Lodge. There, a series of glass houses on the most approved plan, and rows of frames, have been erected. The unemployed have found work by excavating the ground to the depth of some eight feet, and the gravel taken out has made the wide walk across the Green Park and the alterations in the "Mall." A wall and bank of shrubs and trees so completely hides even the highest house in which the palms—such as those outside the National Gallery—are stored, that it is quite invisible from the outside. There are storehouses for the bulbs, and nurseries where masses of wall-flowers, delphiniums, and all the hardier bedding plants, and those for the herbaceous borders, are grown. Of late years the number of beds in the Park has been considerably reduced, without any diminution of the effect. In 1903 as many as ninety were done away with between Grosvenor Gate and Marble Arch. There is now a single row of long beds instead of three rows with round ones at intervals. But even after all these reductions the area of flower beds and borders is very considerable, as the following table will show :—

	Area of Flower Beds.	Area of Flower Borders.
	Sq. Yds.	Sq. Yds.
Hyde Park	1742	2975
Kensington Gardens . .	345	3564
St. James's Park . . .	30	2642
Queue Victoria Memorial in front of Buckingham Palace	1270	...
Total . . .	3687	9181

An event of historic importance which took place in Hyde Park was the Great Exhibition of 1851. Various sites, such as Battersea, Regent's Park, Somerset House, and Leicester Square, were suggested, and the one chosen met with some opposition, but finally the space between Rotten Row and Knightsbridge Barracks was decided on. Plans were submitted for competition, and though 245 were sent in not one satisfied the committee, so, assisted by three well-known architects, they evolved a plan of their own. This was to be carried out in brick ; the labour of removing it after the Exhibition would have been stupendous. It was when this plan was under consideration that Paxton showed his idea for the building of iron and glass so well known as the Crystal Palace. It was 1851 feet long and 408 wide, with a projection on the north 936 feet by 48, and the building covered about 19 acres.

One stipulation was made before the design was accepted, and that was that three great elm trees growing on the site should not be removed, but included in the building. To effect this, some alterations were made, and the trees were successfully encased in this Crystal Palace, and the old trunk of one of them is still standing in Hyde Park. There is a railing round it, but no tablet to record this strange chapter in its history. Some smaller trees were cut down, which led to a cartoon in *Punch* and lines on the Prince Consort, who was the prime mover in all pertaining to the Great Exhibition.

> " Albert ! spare those trees,
> Mind where you fix your show ;
> For mercy's sake, don't, please,
> Go spoiling Rotten Row."

The Exhibition was opened by the Queen on May 1st. The enthusiasm it created in all sections of the population has known no parallel, and in the success and excitement the few small elm trees were soon forgotten by the delighted people, who raised cheers and shouted—

> " Huzza for the Crystal Palace,
> And the world's great National Fair."

Hyde Park never saw more people than during the time it was open from the 1st of May to the 11th of October, as 6,063,986 persons visited the Exhibition,

D

an average of 43,000 daily. Its success was pheno-
menal also from a financial point of view, as after all
expenses were deducted there was a surplus of £150,000,
with which the land from the Park to South Kensington
was purchased, on which the Albert Hall and museums
have been built.

It seems to have been the complete originality of
the whole structure that captivated all beholders. In
his memoirs the eighth Duke of Argyll refers to the
opening as the most beautiful spectacle he had ever
seen. "Merely," he writes, "as a spectacle of joy and
of supreme beauty, the opening of the Great Exhibition
of 1851 stands in my memory as a thing unapproachable
and alone. This supreme beauty was mainly in the
building, not in its contents, nor even in the brilliant
and happy throng that filled it. The sight was a new
sensation, as if Fancy had been suddenly unveiled.
Nothing like it had ever been seen before—its light-
someness, its loftiness, its interminable vistas, its aisles
and domes of shining and brilliant colouring."

It was with the recollection of this world-famous
Exhibition fresh in men's minds that the site for the
Albert Memorial was chosen. The idea conceived by
Sir Gilbert Scott was the reproduction on a large scale
of a mediæval shrine or reliquary. When it was erected
an alteration was made in some of the avenues in Kensington Gardens, so as to bring one into line with the
Memorial. A fresh avenue of elms and planes straight
to the monument was planted, which joined into the
original one, and a few trees were dotted about to break
the old line. As first planned, the avenue must have
commanded a view of Paddington Church steeple in
the vista.

There is no better refutation of the theory that only plane trees will live in London, than an examination of the trees in Hyde Park and Kensington Gardens. An appendix to this volume gives a list of the trees and shrubs which have been planted there, and notes those which are not in existence, having proved unsuitable to London, or been removed from some other cause. Many people will doubtless be surprised at the length of the list. A large number of the trees are really fine specimens, and would do credit to any park in the kingdom. Take, for instance, some of the ash trees. There is a very fine group not very far from the Mount Gate inside Kensington Gardens. Two specimens with light feathery foliage, *Fraxinus lentiscifolia* and *F. excelsior angustifolia*, when seen like lace against the sky, are remarkably pretty trees. Not far from them stand a good tulip tree and the last remaining of the old Scotch firs. The Ailanthus Avenue from the Serpentine Bridge towards Rotten Row, planted in 1876, is looking most prosperous. There are a few magnificent ancient sweet chestnuts above the bastion near the Magazine. The trees planted from time to time have wisely been grouped together according to species. Near the Ranger's Lodge, outside the new frame-ground, some birches grow well, and their white stems are washed every year. The collection of pavias, which flower delightfully in the small three-cornered enclosure where the road divides at the Magazine, are most flourishing. To the south-west of the fountains at the end of the Serpentine, some very good Turkey and American oaks are growing into large trees. Several really old thorns are dotted about. In a walk from the " Round Pond," by the stone which marks the boundary of three parishes, towards Bayswater, grand specimens of oak, ash, lime, elm, sweet and horse-chestnuts are met with. The avenue of horse-chestnuts is just as flourishing as those of planes or elms. In fact the whole Park shows how well trees will succeed if sufficient care is taken of them. One feature of the Park in old days was the Walnut Avenue, which grew nearly on the lines of the present trees between Grosvenor Gate and the Achilles Statue. They were decayed and were cut down in 1811, and the best of the wood was used for gunstocks for the army. It is a pity no walnut avenue was planted instead, as by now it would have been a fine shady walk. The old elms, which are of such great beauty in Hyde Park, have, alas! often to be sacrificed for the safety of pas- sers-by, so that the recent severe lopping was necessary. Their great branches are the first to fall in a gale. Yet when one has to be removed there is an outcry, though people tamely submit to a whole row of trees being ruined by tram lines along the Embankment, so inconsistent is public opinion. It is almost incredible what narrow escapes from destruction even the beauty of Hyde Park has had. In 1884 a Metropolitan and Parks Railway Bill was before Parliament, which actually proposed to cross the Park by tunnels and cuttings which would have completely disfigured "The Dell" and other parts of the Park. In this utilitarian age nothing is sacred.

The Dell had not been ten years in its present form when the proposal was made. The site of the Dell was a receiving lake, about 200 yards by 70, which had been made in 1734. This was done away with in 1844, and the overflow of the Serpentine allowed to pass over the artificial rocks which still remain. It was enveloped in a dark and dirty shrubbery, the haunt of all the ruffians and the worst characters who frequented the Park at night. The place was not safe to pass after dark, neither had it any beauty to recommend it. It was in this state when the present Lord Redesdale became Secretary of the Office of Works in 1874. He conceived the idea of turning it into a subtropical garden, designed the banks of the little stream, and introduced suitable planting, banishing the old shrubs, and merely using the best to form a background to the spireas, iris, giant coltsfoot, osmundas, day lilies, and suchlike, which adorned the water's edge in front. The dark history of the Dell is quite forgotten, and watching the ducks and rabbits playing about this pretty spot is one of the chief delights of Hyde Park.

The monolith which stands near was brought from Liskeard in Cornwall by Mr. Cowper Temple, when First Commissioner of Works, and set up in its present place as a drinking-fountain in 1862. In 1887 the water was cut off it, the railings altered, and the turf laid round it, joining it on to the rest of the Dell. To Lord Redesdale are due also the rhododendrons which make such a glorious show on either side of Rotten Row. He contracted with Messrs. Anthony Waterer for a yearly supply, as they only look their best for a short time exposed to London air. In his time, too, many of the small flower-beds which were dotted about without much rhyme or reason were done away with, and the borders at the edge of the shrubs substituted.

The latest addition to Hyde Park is the fountain presented by Sir Walter Palmer and put up near the end of the "Row" in 1906. The sculpture and design

are the work of Countess Feodore Gleichen. The graceful figure of Artemis, with bow and arrow, and the supporting cariatides, are of bronze, the upper basin of
Saravezza marble, and the lower of Tecovertino stone.
The whole is most light and elegant, and shows up well
against the dark trees.

It has only been possible to glance at the history and
beauties of Hyde Park; many more pages could be
written without touching on half of the incidents connected with it, between the days when it was monastic
lands to the days of the modern Sunday "Church Parade."
It is interesting to trace the origin of the little customs
with which every one is now familiar, but which once
were new and original. For instance, the naming of
trees and flowers in the Parks was first done about 1842,
the idea having been suggested by Loudon, and carried
out by Nash the architect, and George Don the botanist.
Then the system of paying a penny for a seat began in
1820, but when some of the free seats were removed in
1859 there was a great outcry, and they were immediately
put back. Then the meets of the Four-in-hand and
Coaching Clubs, which are quite an institution in Hyde
Park, only continue the tradition of the " Whip Club,"
which first met in 1808. The history of the various
gates calls for notice. The Marble Arch, designed by
Nash, with ornaments by Flaxman, Westmacott, and
Rossi, in Carrara marble, was moved from Buckingham
Palace to its present position in 1851. Over £4000 was
expended on the removal, while the original sum spent
was £75,000. The statue of George IV. by Chantrey,
now in Pall Mall East, was intended for the top, and cost
9000 guineas, and the bronze gates are by Samuel Parker.
Near that corner of the Park was a stone where soldiers

FOUNTAIN BY COUNTESS FEODOR GLEICHEN,
HYDE PARK

were shot, and one of the historians of the Park states that it is still there, only covered over with earth when the new Cumberland Gate was made in 1822. Apsley Gate at Hyde Park Corner was designed by Decimus Burton, and put up in 1827, and he planned the arch forming the entrance to Constitution Hill the following year. The stags, by Bartolozzi, on Albert Gate, came from the Ranger's Lodge in Green Park. Grosvenor Gate was opened about 1724, and Stanhope Gate some twenty-five years later. All the others are more modern.

Those who wish to pursue the subject further will find such details more or less accessible in various guide-books. But to every one the Park, with all its charms, its beauties, and its memories, is open, and it is certain that the better it is known the more it will be appreciated.

CHAPTER III

ST. JAMES'S AND GREEN PARKS

Near this my Muse, what most delights her, sees
A living Gallery of Aged Trees :
Bold sons of Earth, that thrust their Arms so high,
As if once more they would invade the Sky.

. . . .

Here Charles contrives the ord'ring of his States ;
Here he resolves his neighb'ring Princes' Fates ;

. . . . . .

A Prince on whom such diff'rent Lights did smile,
Born the divided World to reconcile.
Whatever Heav'n or high extracted Blood
Could promise or foretel, he'll make it good,
Reform these Nations, and improve them more
Than this fair Park, from what it was before.
—ST. JAMES'S PARK: "Poetical Essay," by Waller.

HE opening history of St. James's and Green Parks is similar to that of Hyde Park. They formed part of the same manor in early days, and became Crown property in Henry VIII.'s time. St. James's Park was chiefly a marsh. The Thames overflowed its banks nearly every year, and the low-lying parts were a swamp and the haunt of wild fowl, and the chief use of the Park was for the sport the wild birds afforded. The Tyburn flowed through it on its way from where it

56

crossed the modern Oxford Street to where it joined the Thames, a little west of where Vauxhall Bridge afterwards stood. It passed right across Green Park, where the depression of its valley can still be traced between Half Moon Street and Down Street. The name, St. James's, originated with the hospital for lepers, dedicated to St. James, on the site of the present palace. The exact date of its foundation is lost in the mists of antiquity, but it was established by the citizens of London, "before the time of any man's memorie, for 14 Sisters, maydens, that were leprous, living chastly and honestly in Divine Service." Later, there were further gifts of land and money from the citizens, and "8 brethren to minister Divine Service there" were added to the foundation. All these gifts were subsequently confirmed by Edward I., who granted a fair to be held for seven days, commencing on the eve of St. James's Day, in St. James's Fields, which belonged to the hospital. The letting out of the land for booths became a source of further income to the lepers. Stowe shortly tells the subsequent history. "This Hospital was surrendered to Henry the 8 the 23 of his reigne : the Sisters being compounded with were allowed Pensions for terme of their lives, and the King builded there a goodly Mannor, annexing thereunto a Park, closed about with a wall of brick, now called St. James's Parke, serving indifferently to the said Mannor, and to the Mannor or Palace of Whitehall." At first sight the summary ejection of these helpless creatures appears unusually heartless, even for those days ; but leprosy, which during the time of the Crusades had grown to a formidable extent, was declining in the sixteenth century in England. It is probable, therefore, that the poor outcast

sisters, possessed of their pensions, would be able to find
shelter in one of the other leper hospitals, of which there
were still a number in the country.

The space between Whitehall and Westminster,
acquired from the Abbey, was turned into an orchard.
The site of Montagu House was the bowling-green of
the Palace, which stretched to the river. A high terrace
and flight of steps led to the Privy Garden of Whitehall,
so, except for the Palace and the Westminster group,
there were no buildings between the river and the Park.
It requires some stretch of the imagination to efface the
well-known edifices which now surround it, and to see it
in its natural state. Flights of wild birds would pass
from the marshy ground to the river, unchecked by the
pile of Government offices. Behind the Leper Hospital
lay fields and scattered houses. The far-off villages of
Knightsbridge and Chelsea would scarcely come into sight,
while beyond the village of Charing the walls and towers
of the City would loom in the distance. Henry VIII.
made some alterations, and may have partially drained
the ground and stocked it with deer. Old maps show
a pond at the west end, near the present Wellington
Barracks, called Rosamund's Pond. The origin of the
name is uncertain, but "Rosemonsbore, or Rosamund's
Bower," occurs in a lease of land near this spot from the
Abbey of Westminster as early as 1520. Hard by was a
"mount," such as was to be seen in every sixteenth-
century garden, probably with an arbour and seat on the
top to overlook the pond. The first mention of St.
James's as a Park is in 1539, on an occasion described
in Hall's Chronicle, when Henry VIII. held a review
of the city militia. "The King himself," writes the
chronicler, "would see the people of the Citie muster

in sufficient nombre. . . ."   Some 15,000, leaving the City after passing by St. Paul's Churchyard, went " directly to Westminster and so through the Sanctuary and round about the Park of St. James, and so up into the fields and came home through Holborne."

It was not until James I.'s time that the Park began to be esteemed as a resort for those attached to the Court. Prince Henry, the elder brother of Charles I., made the tilting-ring on the site of the present Horse Guards' Parade, and brought the enclosure more into vogue for games.   James I. made use of the Park for his own hobbies, one of which was the encouragement of growing vines and mulberries in England.   He planted considerable vineyards, and in 1609 he sent a circular letter to the Lords-Lieutenant of each county, ordering them to announce that the following spring a thousand mulberry trees would be sent to each county town, and people were required to buy them at the rate of three-farthings a plant.   To further prosecute his plan, the King set an example by planting a mulberry orchard at the end of St. James's Park.   The place afterwards became a fashionable tea garden, and Buckingham Palace is partly built on the site.   The King kept also quite a large menagerie of beasts and birds presented to him by various crowned heads, or sent to him by friends and favourites.   There are records of elephants, camels, antelopes, beavers, crocodiles, wild boars, and sables, besides many kinds of birds.   The keepers of the animals received large salaries, and the cost of the care of these beasts would frighten the Zoological Society of to-day.   No expense was spared to give the best and most suitable surroundings to the animals.   For instance, as much as £286 was expended in 1618 by Robert

Wood, the keeper of the cormorants, ospreys, and otters, "in building a place to keep the said cormorants in and making nine fish-ponds on land within the vine garden at Westminster." Fish were put in for these creatures, and a sluice was made to bring water from the Thames to fill the ponds. These strange beasts and birds and their attendants must have been a quaint and unusual sight. The keepers were dressed in red cloth (which cost nine shillings a yard), embroidered with "I.R." in Venice gold, and must have added to the picturesque appearance of this early Zoological Garden.

Gradually the Park became more and more a favourite place in which to stroll. Others were admitted besides the Court circle, the privilege being first accorded to the tenants of the houses at Westminster. Milton, who lived at one time in Petty France, near where Queen Anne's Gate now stands, planted a tree in the garden overlooking the Park, which survived until recent times, would be one of those to enjoy the advantage. Charles I. passed this way on his last journey to Whitehall on the fatal 30th of January, and tradition says he paused to notice a tree planted by his brother Henry. During the Commonwealth, the Park still was resorted to. In the sprightly letters of Dorothy Osborne to Sir William Temple are some vivid little touches in reference to it. She writes from the country in March 1654: "And hark you, can you tell me whether the gentleman that lost a crystal box the 1st of February in St. James's Park or Old Spring Gardens has found it again or not? I have a strong curiosity to know." Again, in June of the same year, she writes from London, where she was paying a visit: "I'll swear they will not allow me time for anything; and to show how absolutely I am governed, I need but tell you that I am every night in the Park and at New Spring Gardens, where, though I come with a mask, I cannot escape being known nor my conversation being admired."

The most brilliant days of its history began, however, in Charles II.'s reign. He entirely remodelled it, and began the work soon after his return from exile, imbued with foreign ideas of gardening. It has always been supposed that Le Nôtre was responsible for the designs, and it has often been asserted that he himself came to England to see them carried out. But close investigation has furnished no proof of this, and it is practically certain that, although invited, and allowed by Louis XIV. to come to England, he never actually did so. Other "French gardeners" certainly came, and one of them, La Quintinge, made many English friends, and kept up a correspondence with them after his return to France. Perrault probably visited London also, and may have superintended the "French gardeners" who were employed on St. James's Park. They transformed the whole place. Avenues—the Mall and "Birdcage Walk"—were planted. A straight canal passed down the middle, and at the end, near the present Foreign Office, was the duck decoy. The "Birdcage Walk" is no fantastic title, for birds were literally kept there in cages. These were probably aviaries for large birds, and not little hanging cages, as has been sometimes suggested. A well-known passage occurs in Evelyn's Diary, 1664, where he enumerates some of the birds and beasts he saw during one of his walks through the Park. The pelican delighted him, although "a melancholy waterfowl," and he watched the skilful way

it devoured fish; and it is not surprising that he recorded the strange fact that one of the two Balearian
cranes had a wooden leg, made by a soldier, with a
joint, so that the bird could "walk and use it as well
as if it had been natural"; and he speaks with interest
of a solan goose, a stork, a milk-white raven, and "a
curious sort of poultry," besides "deer of several
countries," antelopes, elk, "Guinea goats, Arabian
sheep, etc." The duck decoy lay at the south-west end
of the long canal, which formed part of the new French
design. This "duck island" was rather a series of
small islands, as it was intersected by canals and reed-
covered channels for catching duck. This was a
favourite resort of Charles II., who has often been described feeding his ducks in St. James's Park. To be
keeper of the ducks, or "Governor of Duck Island,"
was granted to St. Evremond, an excuse for bestowing
a yearly salary on a favourite. The birds continued
after the King, who had found in them a special recreation, had passed away. In William III.'s time the
Park is still described as "full of very fine walkes and
rowes of trees, ponds, and curious birds, Deer, and some
fine Cows." A Dutch traveller who was in England
from 1693–96 notices the famous old white raven. By
that time the ducks were no longer the fashion, and
evidently there was an inclination to despise the former
craze for wild fowl. A Frenchman, named M. de
Sorbiere, visited England about this time, and wrote an
account of his impressions. Some of his adverse criticisms
of English people and institutions got him into trouble.
A supposed translation of his book was published in
1698, and until 1709 was held to be a correct version.
In reality it was a clever skit, and not in the least like

the original. In the true version he describes the Park with its rows of trees and "admirable prospect" of the suburbs, and mentions that the King had "erected a tall Pile in the Park, the better to make use of Telescopes, with which Sir Robert Murray shew'd me Saturn and the Satellites of Jupiter." Not a word about the ducks. But in the spurious parody of 1698 there is a humorous description, which shows how the next generation laughed at the amusements of King Charles II. "I was at St. James's Park; there were no Pavillions, nor decoration of Treilliage and Flowers; but I saw there a vast number of Ducks; these were a most surprising sight. I could not forbear to say to Mr. Johnson, who was pleased to accompany me in this Walk, that sure all the ponds in England had contributed to this profussion of Ducks; which he took so well, that he ran immediately to an Old Gentleman that sate in a Chair, and was feeding of 'em. He rose up very obligingly, embraced me, and saluted me with a Kiss, and invited me to Dinner; telling me he was infinitely oblig'd to me for flattering the King's Ducks."

Little attention was paid to the wild fowl in the Park after that date, until the Prince Consort took an interest in them. In 1841 he became the Patron of the Ornithological Society, and the cottage on Duck Island was built for the Bird-keeper. For some thirty years the Society flourished, and kept up the supply and cared for the birds in the Park. In 1867, however, their numbers were greatly reduced, and the Society sold their collection of birds to H.M. Office of Works, which has since then had them under its charge. It is pleasant to know that the old tradition of the wild fowl in that part of the Park is maintained. Although the duck pond of King Charles's time must have looked somewhat different from that of to-day, the birds can be made as much at home, and they nest peacefully on the modern Duck Island, its direct descendant. Moorhens and dabchicks, or little grebes, have for the last twenty years nested in the Park. They used to leave for the breeding season, but since 1883, when the first moorhen nested, they have gradually taken to remaining contentedly all through the year, and bring up their young there. Birds seem to choose the Park to rest in, and many migratory ones have been noticed. Kingfishers have recently been let out near the site of the ancient bird cages, in the hope that they may carry on the historic association.

The cows, which were a part of ancient history, as were the birds, have not been so fortunate. Although a newspaper clamour in defence of the cows was raised, the few remaining were finally banished in 1905, when the alterations in the Mall were made. These survivals standing by the dusty stalls could scarcely be called picturesque; and although interest undoubtedly was attached to them as venerable survivals of an old custom, they hardly suggested the rural simplicity of the days when cows were really pastured in the Park. For over two centuries grazing was let to the milk-women who sold milk at the end of the Park, near Whitehall. They paid half-a-crown a week, and after 1772 three shillings a week, for the right to feed cattle in the Park. A Frenchman, describing St James's at that time, is astonished at its rural aspect. "In that part nearest Westminster nature appears in all its rustic simplicity; it is a meadow, regularly intersected and watered by canals, and with willows and poplars, without any regard to

CROCUSES IN EARLY SPRING, ST. JAMES'S PARK

order. On this side, as well as on that towards St. James's Palace, the grass plots are covered with cows and deer, where they graze or chew the cud, some standing, some lying down upon the grass. . . . Agreeably to this rural simplicity, most of these cows are driven, about noon and evening, to the gate which leads from the Park to the quarter of Whitehall. Tied to posts at the extremity of the grass plots, they swill passengers with their milk, which, being drawn from their udders on the spot, is served, with all cleanliness peculiar to the English, in little mugs at the rate of a penny a mug." The combination of the gay crowd in hooped petticoats, brilliant coats, and powdered wigs, with the peaceful, green meadows and the browsing deer and cows, forms an attractive picture.

All this had changed long before the final departure of the cattle, when the last old woman was pensioned off, and the sheds carted away. A use was found for the fragments of the concrete foundations of the last milkmaid's stall. They were made into a sort of rockery, on which Alpine plants grow well, to support the bank at the entrance to the new frame-grounds at Hyde Park.

But to return to Charles II.'s time, when the cows were undisturbed. The great feature of what Pepys calls the "brave alterations" was the canal. He mentions more than one visit when the works were in progress. In October 1660 he went "to walk in St. James's Park, where we observed the several engines at work to draw up water, with which sight I was very much pleased." The canal, when finished, was 2800 feet long and 100 broad, and ran through the centre of the Park, beginning near the north end of Rosamund's Pond. An avenue of trees was planted on either side,

E

passing down between the canal and the duck decoy to
a semicircular double avenue near the tilting-ground.
Deer wandered under fine old oaks between the canal
and the avenues of " the Mall." These old trees have
gradually disappeared, as much through gales as from
the wanton destruction of the would-be improver. At
the hour of Cromwell's death, when the storm was so
fierce the Royalists said it was due to fiends coming to
claim their own, much havoc was wrought; and from
time to time similar destructions have taken place, one
of the most serious being in November 1703, when
part of the wall and over 100 elms were blown down.
Another notable gale was on March 15, 1752, when
many people lost their lives. " In St. James's Park and
the villages about the metropolis great numbers of trees
were demolished."

The broad pathway, between avenues on the opposite
side of the Park to the Birdcage Walk, now called the
Mall, derives this name from the game of " paille-
maille," which is known to have been played in France
as early as the thirteenth century, and which was popular
in England in the seventeenth. The locality, however,
where it was first played in James I.'s time was on the
northern side of the street, which is still called from it,
Pall Mall. In those days fields stretched away beyond
where now St. James's Square lies, and a single row of
houses lay between the playground and the Park. As
the game became more the fashion, the coaches and dust
were found too disturbing for enjoyment, and a new
ground was laid out, running parallel to the old one,
but within the Park. The game is considered by some
to be a forerunner of croquet, as it was played with a ball
(=*pila*) and mallet, the name being derived from these

two words.   One or more hoops had to be passed
through, and a peg at the further end touched.   The
winner was the player who passed the hoops and reached
the peg in the fewest number of strokes.   The whole
course measured over 600 yards, and was kept brushed
and smooth, and the ground prepared by coating the
earth with crushed shells, which, however, remarked
Pepys, " in dry weather turns to dust and deads the
ball."   Both Charles II. and James II. were much
addicted to the game, and the flattering poet Waller
eulogises King Charles's " matchless " skill :—

> " No sooner has he touched the flying ball,
>   But 'tis already more than half the Mall."

The Park was by his time a much-frequented spot,
and crowds delighted to watch the King and his courtiers
displaying their dexterity.   Charles II. is more intimately
connected with St. James's Park than any other great
personage.   He sauntered about, fed his ducks, played
his games, and made love to fair ladies, all with indulgent, friendly crowds watching.   He stood in the
" Green Walk," beneath the trees, to talk with Nell
Gwynn, in her garden " on a terrace on the top of the
wall " overlooking the Park ; and shocked John Evelyn,
who records, in his journal, that he heard and saw " a
very familiar discourse between the King and Mrs.
Nelly."   Charles's well-known reply to his brother, that
no one would ever kill him to put James on the throne,
was said in answer to James's protest that he should not
venture to roam about so much without attendants in
the Park.   His dogs often accompanied him, and
perhaps, like most of their descendants, these pets had
a sporting instinct, and ran off to chase the deer.   Any-

how, they managed frequently to escape their master's vigilance, and fell a prey to the unscrupulous thief, and descriptions of the missing dogs were published in the Gazette.   One, answering to the name Towser, was "liver colour'd and white spotted"; and a "dogg of His Majestie's, full of blew spots, with a white cross on his forehead about the bigness of a tumbler," was lost on another occasion.

Charles with his dogs, his ducks, his wit, his engaging manners, his doubtful morals, is the central figure of many a picture in St. James's Park, but it does not often form a background to his Queen.   One scene described by Pepys has much charm.   The party, returning from Hyde Park on horseback with a great crowd of gallants, pass down the Mall; the Queen, riding hand in hand with the King, looking "mighty pretty" in her white laced coat and crimson petticoat.   Again, on another occasion, the Queen forms an attractive vision, as she walks with her ladies from Whitehall to St. James's dressed from head to foot in silver lace, each holding an immense green fan to shade themselves from the fierce rays of the June sun, while a delighted crowd throng round them.

The popularity of the Mall as the rendezvous of all classes lasted for over a century.   Through the reigns of Queen Anne and George I. and II. all the fashionable world of London congregated there twice daily.   In the morning the promenade took them there from twelve to two, and after dinner in full dress they thronged thither again, not to play the game of paille-maille, which was then out of fashion, but simply to walk about under the trees and be amused with races, wrestlings, or an impromptu dance.   Every well-known person—courtiers, wits, beaux, writers, poets, artists, soldiers—and all the beautiful and fascinating women, great ladies as well as more humble charmers, and bold adventuresses, were to be seen there daily.

The crowds seem to have been very free in their admiration of some of the distinguished ladies. When the three lovely Misses Gunning captivated everybody with their wit and beauty, they had only to appear in the Mall to be surrounded by admirers. On one occasion they were so pressed by the curious mob that one of these matchless young charmers fainted and had to be "carried home in a sedan."

On looking at an old print of the ladies in their thin dresses walking in the Mall, it is customary to bemoan the change of climate, to wonder if our great-great-grand-mothers were supernaturally strong and not sensitive to cold, or to conclude that they only paraded there in fine weather. Apparently this last is not the correct solution, for in 1765 they astonished Monsieur Grosley by their disregard of the elements. He is horrified at the fog. "The smoke," he writes, "forms a cloud which envelopes London like a mantle; a cloud which the sun pervades but rarely; a cloud which, recoiling back upon itself, suffers the sun to break out only now and then, which casual appearance procures the Londoners a few of what they call *glorious days*. The great love of the English for walking defies the badness of other days. On the 26th April, St. James's Park, incessantly covered with fogs, smoke, and rain, that scarce left a possibility of distinguishing objects at a distance of four steps, was filled with walkers, who were an object of musing and admiration to me during the whole day." Few ladies nowadays fear a little fog or rain, but to walk in it they must be attired in short skirts, thick boots, and warm or mackintosh coats. It must have been much more distressing in the days of powdered hair, picture hats, and flimsy garments. No wonder M. Grosley was astounded at the persistence of the poor draggled ladies.

All foreign visitors to London naturally went to see the Mall. Here is the account of a German baron, describing the man of the world : " He rises late, dresses himself in a frock (close-fitting garment, without pockets, and with narrow sleeves), leaves his sword at home, takes his cane, and goes where he likes. Generally he takes his promenade in the Park, for that is the exchange for the men of quality. 'Tis such another place as the Garden of the Tuileries in Paris, only the Park has a certain beauty of simplicity which cannot be described. The grand walk is called the Mall. It is full of people at all hours of the day, but especially in the morning and evening, when their Majesties often walk there, with the royal family, who are attended only by half-a-dozen Yeomen of the Guard, and permit all persons to walk at the same time with them."

A writer in 1727, waxing eloquent on the charms of the Park, gives up the task of describing it, as " the beauty of the Mall in summer is almost past description." "What can be more glorious than to view the body of the nobility of our three kingdoms in so short a compass, especially when freed from mixed crowds of saucy fops and city gentry?" But more often the company was very mixed, and manners peculiar. This brilliant and motley assembly indulged in all kinds of amusements. Even the grandest frequenters afforded diversion some-times to the " saucy fops." Wrestling matches between various courtiers attracted crowds, or a race such as one between the Duke of Grafton and Dr. Garth, of 200 yards, was the excitement of the day. There were odd and original races got up, and wagers freely staked. Some inhuman parents backed their baby of eighteen months old to walk the whole length of the Mall (half a mile) in thirty minutes, and the poor little mite performed the feat in twenty-three minutes. What comments would modern philanthropic societies have made on such a performance!

A race between a fat cook and a lean footman caused great merriment, but as the footman was handicapped by carrying 110 lbs., the fat cook won. Another time it was a hopping-race which engrossed attention—a man undertook to hop one hundred yards in fifty hops, and succeeded in doing it in forty-six—and endless variety of similar follies. The crowds who assembled indulged in every sort of gaiety; "in short, no freedoms that can be taken here are reckoned indecent; all passes for raillery and harmless gallantry."

Although open to all the world for walking, only royal personages or a few specially favoured people were allowed to drive through. It was one of the grievances of the Duchess of Marlborough when the Duke was in disgrace that the privilege of driving her coach and six through the Park was denied her. The remaining restrictions with regard to carriages have only passed away in very recent years. The notice board stating that Members of Parliament during the session might drive through the Park from Great George Street to Marlborough House was only removed when the road was opened to all traffic in 1887, and Constitution Hill only became a public highway in 1889. The use of the road passing under the Horse Guards' Archway is still restricted to those who receive special permission from the sovereign.

The Park had never been drained, and had always shown signs of its marshy origin, and "Duck Island" was really a natural swamp. An unusually high tide flooded the low-lying end where the Horse Guards' Parade and the houses of Downing Street with their little gardens now stand. What state secrets they could divulge had they the power of speech! The tilting-ground was often in a condition quite unfit for the exercise of troops, so with a view to preventing this, it was paved with stone early in the eighteenth century. It has always been used for military displays, and the trooping of the colours on the King's birthday takes place on the same ground which witnessed the brilliant scene when the colours, thirty-eight in number, captured at the battle of Blenheim were conveyed to Westminster Abbey. On the parade-ground now stands the gun cast at Seville, used by Soult at Cadiz, and taken after the battle of Salamanca. Here many an impressive cere-mony of distributing medals, and countless parades, have taken place through many generations. Here, with the brutality of old days, corporal punishment was ad-ministered, and offending soldiers were flogged in full view of the merry-making crowds assembled in the Park. Round the Park lay other marshy lands, also frequently flooded by the Thames, and it was not surprising that on one occasion an otter found its way from the river and settled down on Duck Island and there grew fat on the King's carp. Sir Robert Walpole sent to Houghton for his otter-hounds, and an exciting hunt ensued, in which the Duke of Cumberland took part, and the offending otter was captured.

Rosamund's Pond had, in the course of time, become stagnant and unpleasant, and there were frequent complaints of its unsavoury condition. About 1736 a machine for pumping out water was invented by a Welshman, and used successfully to empty the pond, and it was thoroughly cleansed. Thirty years later the same evil began again to be a nuisance, and it was decided to drain and fill up the pond entirely, which was accomplished about 1772. The trees on the island were felled, and those near the bank died from the lack of water, so at first the absence of the slimy pond must have been disfiguring. The shady walk near it, known as the Close Walk or the Jacobites' Walk, must have disappeared when the trees died. About the same time the swampy moat round Duck Island was filled up and the canal cleaned out. When these improvements were completed in 1775 some birds were put on the canal. One of them was a swan called Jack, belonging to Queen Charlotte, which was reared in the garden of Buckingham House. This bird ruled the roost for many a day, and was a popular favourite. It lived until 1840, when some new arrivals, in the shape of Polish geese, pecked and ill-treated the poor old bird so seriously that he died.

About 1786 fashion began to desert the Mall for the Green Park, and the crowds which collected there were no longer intermingled with the Court circle. In a letter to her daughter Madame Roland describes the company in the Mall as very different from what it was a few years earlier, for though it was "very brilliant on a Sunday evening, and full of well-to-do people and well-dressed women, in general they are all tradespeople and citizens." A generation later the Mall seems to have become quite deserted. Sir Richard Phillips, in his morning's walk from London to Kew in 1817, bemoans the absence of the gay throng :—

"My spirits sank, and a tear started into my eyes, as I brought to mind those crowds of beauty, rank, and fashion which, until within these few years, used to be displayed in the centre Mall of this Park on Sunday evenings during spring and summer. How often in my youth had I been the delighted spectator of the enchanted and enchanting assemblage. Here used to promenade, for one or two hours after dinner, the whole British world of gaiety, beauty, and splendour. Here could be seen in one moving mass, extending the whole length of the Mall, 5000 of the most lovely women in this country of female beauty, all splendidly attired, and accompanied by as many well-dressed men. What a change, I exclaimed, has a few years wrought in these once happy and cheerful personages! How many of those who on this very spot then delighted my eyes are now mouldering in the silent grave!"

About 1730 Queen Caroline, who was then busy with the alterations in Hyde Park, turned her attention to what is now known as the Green Park also. It had all formed part of St. James's Park, and was known as the Upper Park or Little St. James's Park. It was enclosed by a brick wall in 1667 by Charles II., who stocked it with deer. In the centre of the Park an ice-house was made, at that time a great novelty in this country, although well known in France and Italy. In his poem on St. James's Park Waller alludes to it :—

"Yonder the harvest of cold months laid up
  Gives a fresh coolness to the royal cup ;
There ice like crystal firm and never lost
  Tempers hot July with December's frost."

No further alterations were made, except that, in 1681, Charles effected an exchange of land with the Earl of Arlington, on which, a few years later, Arlington Street was built. The path which runs parallel with the backs of these houses was Queen Caroline's idea, and she used it frequently herself, and it became known as the "Queen's Walk." The houses overlooking the Park went up in value as the occupants could enjoy the sight of the Queen and the Princesses taking their daily walk. The line of this path is no longer the same, as a piece was cut off the Park in 1795 and leased to the Duke of Bridgewater to add to the garden of his house. The Queen also built a pavilion known as the Queen's Library in the Park, where she spent some time after her morning promenades. Although Queen Caroline took to the Upper Park, the world of fashion did not follow at once, and it was not until about 1786 that the Green Park for some reason suddenly became the rage. The only incident of historic interest between this date and the making of the road was the celebration of the end of the War of Succession in the spring following the Peace of Aix-la-Chapelle. A great pavilion like a Doric temple, 410 feet long and 114 feet high, was erected near the wall separating the Green Park from St. James's, and on the 27th of April a grand display of fireworks was arranged. A fire, however, broke out just as the performance was beginning, when a grand overture composed by Handel had been performed, and the King and dense crowds were watching the illuminations. The flames were got under, but not before much of the temporary building had been destroyed, and the greater part of the fireworks perished in the flames, and several fatal and serious accidents further marred the entertainment.

Near the top of the Park was a reservoir or "fine piece of water" belonging to the Chelsea Waterworks, and the path round it was included in the fashionable promenade by those who paraded in the Queen's Walk after dinner. Lower down, where there is still a depression, was a little pond, originally part of the Tyburn stream. The "green stagnant pool" was abused by a writer in 1731, who regretted that trees had just been planted near it, which probably meant that the offensive pool would "not soon be removed." The prophecy was correct, for it was more than a hundred years later before this was filled up. The Park wall ran along Piccadilly, and here and there, as was often the case in the eighteenth century, there were gaps with iron rails, through which glimpses of the Park could be obtained. Some persons had private keys to the gates leading into the Park from Piccadilly. Daring robberies were by no means uncommon, and thieves, having done mischief in the streets near Piccadilly on more than one occasion, were found to be provided with keys to the gates, through which they could make their escape into the Park and elude their pursuers. The Ranger's Lodge stood on the northern side, and was rebuilt and done up in 1773. It was made so attractive that there was great competition, when it was completed, to be Deputy-ranger and live there. The two stags which now stand on Albert Gate, Hyde Park, once adorned the gates of this Ranger's Lodge. It is described in 1792 as "a very neat lodge surrounded by a shrubbery, which renders it enchantingly rural." When George III. bought Buckingham House, then an old red-brick mansion, he took away the wall which separated the Green Park from St. James's, and put a railing instead. In this wall was another lodge, and a few trees near it, known as the Wilderness.

The aspect of the Mall has greatly changed since the days when its fashion was at its height. Then the gardens of St. James's Palace ran the whole length of the north side from the Palace towards Whitehall. Stephen Switzer, writing in 1715, extols the beauty of the garden, which by his time was cut up and partly built on. "The Royal Garden in St. James's Park, part of which is now in the possession of the Right Honourable Lord Carlton, and the upper part belonging to Marlborough House, was of that King's [Charles II.] planting, which were in the remembrance of most people the finest Lines of Dwarfs perhaps in the Universe. Mr. London" . . . presumed " before Monsieur de la Quintinge, the famous French gardener, . . . to challenge all France with the like, and if France, why not the whole World ? "

Carlton House, a red-brick building, with the stone portico now in front of the National Gallery, was built in 1709 on part of this garden. Some twenty years later, before it was purchased by Frederick, Prince of Wales, the grounds belonging to the house were laid out by Kent. Until Carlton House was pulled down in 1827, therefore, the Mall was bounded on the north by choice gardens. Between the Mall and the walls of these gardens ran the " Green Walk," or " Duke Humphrey's Walk," as it was also often called. The origin of the latter name is to be traced to old St. Paul's. The monument to Humphrey, Duke of Gloucester, in the centre aisle of old St. Paul's Cathedral was where " poore idlers " and " careless mal-contents " congregated—

"Poets of Paules, those of Duke Humfrye's messe
That feed on nought but graves and emptinesse."

When Duke Humphrey's Walk in St. Paul's was burnt the name became attached to the walk in St. James's Park, where idlers also sauntered. Some writers attribute the transference of the name to the fact that the arched walk under the trees was like the cathedral aisle. Anyhow the name clung to this walk in the Park from 1666 and during the eighteenth century.

When Carlton House became the centre of attraction the Park itself was in a very neglected state. The canal was turbid, the grass long, and the seats unpainted. How long it would have remained in this condition is uncertain had not a new impulse of gardening possessed the whole nation, and once more it was resolved to alter the entire Park.

The rage for landscape gardening was at its height. Capability Brown had done his work of destruction, and set the fashion of "copying nature," and his successors were following on his lines, but going much further even than Brown. The sight of a straight canal had become intolerable. The Serpentine was designed when the idea that it might be possible to make the banks of artificial sheets of water in anything but a perfectly straight line was just dawning, but the canal in St. James's Park was transformed when half the stiff ponds and canals in the kingdom had been twisted and turned into lakes or meres. Brown had had a hand in the alterations at the time Rosamund's Pond was removed, but it was Eyton who planned and executed the work fifty years later. It was begun in 1827, and a contemporary writer praises the result as "the best obliteration of avenues" that has been effected. Although he owns it involved "a tremendous destruction of fine elms," he is lost in admiration of the "astounding ingenuity" which "converted a Dutch canal into a fine flowing river, with incurvated banks, terminated at one end by a planted island and at the other by a peninsula." A permanent bridge was first made across the water about this time. Previously a temporary one had been made when the Allied Sovereigns visited London in 1814—a kind of Chinese design by Nash, surmounted by a pagoda of seven storeys. It was this flimsy edifice which made Canova say the thing that struck him most in England was that Waterloo Bridge was the work of a private company, while this bridge was put up by the Government. It was on the canal in St. James's Park that skates of a modern type first appeared in London. Bone ones were in use much earlier on Moorfields. Both Evelyn and Pepys saw the new pattern first in the Park in 1662. Two years later Pepys notes going to the canal with the Duke of York, "where, though the ice was broken and dangerous, yet he would go slide upon his scates, which I did not like, but he slides very well." Just before the alterations began, and the complete change of the canal was taken in hand, the Park was lighted with gas lamps, an innovation which caused much excitement. At the same time orders were issued to shut the gates by ten every evening. A wit on this occasion wrote the following lines, which were found stuck up on a tree :—

> " The trees in the Park
> Are illumined with gas,
> But after it's dark
> No creatures can pass.

> " Ye sensible wights
> Who govern our fates,
> Extinguish your lights
> Or open your gates."

The same lamps inspired another poet, who wrote, just before the destruction of the avenues took place :—

"Hail, Royal Park! what various charms are thine;
Thy patent lamps pale Cynthia's rays outshine,
Thy limes and elms with grace majestic grow
All in a row."

Yet once more has St. James's Park been subjected to renovation. The work, which is a memorial to our late beloved Queen Victoria, is not yet completed, so its description must be imperfect. The design aims at drawing together the several quarters of the Park towards Buckingham Palace and a central group of statuary. The Mall is now the scene of ceaseless traffic, and the sauntering pedestrian is a thing of the past. A wide road runs at right angles across the Green Park, and so once again more closely associates the Upper with the Lower St. James's Park. Probably the greatest praise of the alterations would be to say that Le Nôtre would have approved them. They seem to complete the design in a fitting manner, but they banish once and for all time, the semi-rural character which for so many centuries clung to the Park. The design includes a series of formal parterres which are filled with bedding-out plants raised in Hyde Park. In the summer of 1906 they were planted with scarlet geraniums with an edging of grasses and foliage and a few golden privets, and on hot July days there were many people ready to pronounce the arrangement as extremely bad taste. It seemed a reversion to the days when a startling mass of colour was the only effect aimed at. As they appeared all through the mild October days, when a soft foggy light enveloped the world, and the trees looked dark and

dreary, with their leaves, devoid of autumn tints, still
struggling to hold on, the vivid colouring of the beds
gave a very different impression. The charm of the warm
red tone against the cold blue mists must have given a
sensation of pleasure to any one sensitive to such contrasts.

A CORNER OF THE QUEEN VICTORIA MEMORIAL GARDENS,
IN FRONT OF BUCKINGHAM PALACE

The Park in spring has nothing of the stiff, early
Victorian gardening left. Under the trees crocuses raise
their dainty heads, as cheerily as from out of Alpine
snows, and the slopes of grass spangled with a " host of
golden daffodils " are a delight to all beholders.

F

The palmy days of St. James's Park may have passed away—no longer is the fate of nations and the happiness of lives decided under its ancient elms—but those days have left their mark. Every path, every tree, every green-sward, could tell its story. The Park is now more beautiful than it ever was, even though fashion has deserted it. The last changes are but one more link in the long historic chain. It brings the Park of the Stuarts, the Mall of the Queen Anne's age of letters, down to our own great Queen and the days of Expansion and Empire. A stroll under its shady trees and by its sparkling water must be replete with suggestions to the moralist, with thoughts to the poet, and with an inexpressible charm to the ordinary appreciative Londoner.

CHAPTER IV

REGENT'S PARK

When Philomel begins to sing
The grass grows green and flowers spring;
Methinks it is a pleasant thing
To walk on Primrose Hill.
—Roxburgh Ballads, *c.* 1620.

EGENT'S PARK has had but a
transitory day of fashion, and his-
tory has not crowded it with associations like the other Royal Parks.
It is the largest and one of the most
beautiful, yet there is something
cold and less attractive about it.
In spring, with its wealth of thorn
trees, it has a delightfully rural appearance, and it possesses many charms on close acquaintance. Its history
as a Royal Park is as ancient as that of Hyde Park
or St. James's, but it remained a distant country sporting estate, and only assumed the form of a Park, in
the modern sense of the word, less than a hundred
years ago.

In the dim distance of Domesday it formed part of
the manor of Tybourne. Later on the manor became
Marylebone or Mary le Bourne, the Church of St. Mary
by the Burn, the brook in question being the Tyburn.
The manor in Domesday is described as part of the

83

lands belonging to the Abbey of Barking in Essex.   In
the thirteenth century it was held by Robert de Vere,
and passed by descent through his daughter to the
Earls of Arundel.   Later on the manor was divided,
and a fourth share came to Henry V. as heir to the
Earls of Derby.   The greater part of the manor was
bought by Thomas Hobson, and his son, who was Lord
Mayor in 1544, exchanged it with Henry VIII. for some
church lands elsewhere.   So it became part of the royal
hunting-ground, and the same enactment concerning
the preservation of game applied to Marylebone Park,
situated within the manor, as to Hyde Park.   Queen
Elizabeth leased part of the manor to a certain Edward
Forset, and James I. sold him all the manor except the
part known as Marylebone Park, now Regent's Park.
It was again sold by the grandson of Edward Forset
to John Holles, Duke of Newcastle, and passed to his
daughter, who married Edward Harley, Earl of Oxford,
and through their daughter, who married the second
Earl of Portland, to the Bentinck family.   The Park
has always remained Crown property, although it has
frequently been let by the reigning sovereign.   Charles I.
granted it to Sir G. Strode and J. Wandesford as a
payment of a debt of £2318 for arms and ammuni-
tion.   It was sold by Cromwell with all the other
royal lands, but after the Restoration it went back
to its former holders till the debt was discharged,
and after that to various other tenants.   It was on
the expiration of a lease to the Duke of Portland in
1811 that the laying out of the Park in its present
form commenced.

During the early period incidents connected with
it are meagre.   It is for the most part only in royal

accounts that references to Marylebone Park are found, and they are merely a bare statement of facts. But that hunting-parties, with all the show and splendour attending them, took place frequently, is certain. Among the Loseley MSS. occur, in 1554, instructions to Sir Thomas Cawarden, as "Master of the Tents and Toiles," to superintend the making of "certaine banquiting houses of Bowes [=boughs] and other devices of pleasure." One of these was made in "Marybone Parke," and a minute description is given. It was 40 feet long, and "wrought by tymber, brick, and lyme, with their raunges and other necessary utensyles therto insident, and to the like accustomed." Also three "standinges" were made at the same time, "all of tymber garnished with boughes and flowers, every [one] of them conteynenge in length 10 foote and in bredth 8 foote, which houses and standings were so edified, repaired, garnished, decked, and fynyshed against the Marshall Saint Andrewes comynge thethere by speciale and straight comandement, as well of the late King as his counsell to Sir Thos· Cawarden, Knt. Mr· of the said Office of Revels; and Lawrence Bradshaw, Surveior of the King's works, exhibited for the same wt· earnest charge done, wrought and attended between the 27th of June and the 2 of August in the said year" [4th of Edward VI.]. Employed on the above works for 22 days at all hours, a space to eat and drink excepted, "Carpenters, bricklayers, 1d. the hour; labourers, ½d. p. hour; plasterers, 11d. a day; painters, 7d. and 6d. a day." "Charges for cutting boughs in the wood at Hyde Park for trimming the banquetting house, gathering rushes, flags, and ivy; painters, taylors for sewing roof, etc., basket makers working upon windows, total cost, £169, 7s. 8d." Only about half of this total was due to the work in Marylebone, as a similar pavilion, and three other "standings," were made in Hyde Park at the same time.

Hall, the chronicler of Henry VIII.'s time, inveighs against the fashion of making these sumptuous banqueting houses. They were not only a regal amusement, but the citizens built in their suburban gardens "many faire Summer houses . . . some of them like Midsummer Pageants, with Towers, Turrets, and Chimney tops, not so much for use or profit, as for shew and pleasure, and bewraying the vanity of men's mindes, much unlike to the disposition of the ancient Citizens, who delighted in building of Hospitals and Almeshouses for the poore." There stood in Marylebone parish a banqueting house where the Lord Mayor and aldermen dined when they inspected the conduits of the Tybourne. On one occasion they hunted a hare before dinner, and after, "they went to hunt the fox. There was a great cry for a mile, and at length the hounds killed him at the end of St. Giles." During this run the hunt must have skirted the royal preserves of Marylebone. In Elizabeth's time a hunting-party on 3rd February 1600 is recorded, in which the "Ambassador from the Emperor of Russia and the other Muscovites rode through the City of London to Marylebone Park, and there hunted at their pleasure, and shortly after returned homeward."

Marylebone was a retired spot for duels, and many took place there down to the time when duelling ceased. The quarrel which led to one in Elizabeth's reign is most typical of that age. Sir Charles Blount, afterwards Earl of Devonshire, handsome and dashing, distinguished himself in the lists, and won the approbation of Queen Elizabeth. She presented him with a chessman in gold, which he fastened on his arm with a crimson ribbon. This aroused the jealousy of Essex, who said with scorn, "Now I perceive that every fool must have a favour." Whereupon Blount challenged him. They met in Marylebone Park, and Essex was disarmed and wounded in the thigh.

In Mary's time the Park witnessed a warlike scene in connection with one of the organised attempts to dethrone the Queen. The indictment of Sir Nicholas Throgmorton for high treason, because he, with Sir Thomas Wyatt and others, "conspired to depose and destroy the Queen," states that "the said Sir Nicholas plotted to take and hold the Tower, levy war in Kent, Devonshire, etc., and, with Sir Henry Isley and others, on 26 January 1554, rose with 2000 men, marched from Kent to Southwark, and by Brentford and Marylebone Park to London, the Queen being then at Westminster, but were overthrown by her army." The incidents which centre round this Park are few. Even in the accounts of all the royal lands it does not often occur. In 1607 one item in the Domestic State Papers, a list of nine parks, from each of which four bucks were to be taken, includes Hyde Park, but Marylebone is not mentioned, and in orders to the keepers it does not often occur.

During the Commonwealth it comes more into notice, from the sad fact that it was then sold and disparked, and the trees cut down. When Cromwell sold it to "John Spencer of London, gent.," the proceeds were settled on Col. Thomas Harrison's regiment of dragoons for their pay. The existing Ranger, John

Carey, was turned out, and Sir John Ipsley put in his place. The price given for the Park was £13,215, 6s. 8d., which included £130 for deer and £1774 for timber, exclusive of 2976 trees which were marked for the Royal Navy. Cromwell probably knew the Park and its advantages well, as some years before, when he was a boy, his uncle had had permission to hunt in any of the royal forests. The warrant is dated 15th June 1604, "to the lieutenants, wardens, and keepers of the forests, chases, and parks, to permit Sir Oliver Cromwell, Knt., Gentleman of the Privy Chamber, to hunt where he shall think fit." The work of hewing the timber began at once. On October 19, 1649, the Navy Commissioner was instructed to "repair the crane at Whitehall for boating timber, which is to go from Marylebone Park to the yards to build frigates." Again, Sir Henry Mildmay was ordered to "confer with Mr. Carter, Surveyor of Works, for the timber in Marylebone Park to be brought through Scotland Yard, to be boated there for use of the navy." Cromwell converted the Park to other uses, as in June the same year orders were given to put to grass in Marylebone Park all the artillery horses "bought by Captain Tomlins for Ireland till Monday week." That a number were turned out there for a time is clear from the further warrant, dated July 12, to "permit William Yarvell, Carriage Master, to put all the horses provided for Ireland, which cannot be accommodated in Marylebone Park, into Hyde Park to graze." No doubt they found excellent pasture, in spite of the game. Still, the deer must have been fairly numerous, considering the price paid for those left when the Park was sold. One hundred of the "best deer" were first ordered to be removed from there to St. James's Park, "Colonel Pride to see to the business."

At the Restoration the former tenants were reinstated until the debt was discharged, and John Carey was compensated for his loss of the rangership; but the Park was never re-stocked with deer. It is supposed that the Queens, Mary and Elizabeth, sometimes resided at the Manor House belonging to the Manor, which stood at the south side of what is now Marylebone Road, and was built by Henry VIII. A drawing of the house in 1700 exists, and it is not the same as Oxford House, with which it has sometimes been confused, belonging to Lord Oxford, which contained the celebrated Harleian collection of MSS. Henry VIII.'s Manor House was pulled down in 1790. It is not until after that date that anything further has to be recorded of the Park; until then it remained let out as farms. In 1793 Mr. White, architect to the Duke of Portland, the tenant of the Park from the Crown, approached Mr. Fordyce, the Surveyor-General, with his ideas and plans for the improvement of the whole of the area. During the previous fifty years the streets and squares between Oxford Street and Marylebone had been growing up. Foley House, a large building, stood on the site of the present Langham Hotel; and in the lease by which the land was held from the Duke of Portland, it was covenanted that no buildings should obstruct the view of Marylebone Park from this house. When, in 1772, the Brothers Adam designed Portland Place, they made it the entire width of Foley House, so that the agreement was fulfilled to the letter. In those days the street ended where No. 8 Portland Place now stands; then came the railings which enclosed Marylebone Fields, with its buttercup meadows

and country lanes and hedgerows. White's idea commended itself to Fordyce, and he approached the Treasury
on the subject. The total area, according to the survey
in 1794, was 543 ac. 17 p. This was disposed chiefly
between three farms of about 288, 133, and 117 acres
respectively. From the first all the plans embraced
extensive buildings, as well as a proportion of park.
Inspired by Fordyce, the Treasury offered a prize, not
exceeding £1000, for the best design, and several were
submitted. Fordyce aimed at something between the
most extreme votaries of the landscape school and the
older, debased, formal styles—a compromise which Loudon
was at that time trying to bring into vogue. A " union
of the ancient and modern styles of planting," he called it,
which led by stages to the Italian parterres and brilliant
bedding out of the early Victorian gardens. Fordyce
did not live to see any plan put into execution. At his
death the Surveyor-General of Land Revenues and the
Commissioners of Woods and Forests were amalgamated,
and Leverton and Chawner, architects to the former, and
Nash, architect to the latter, submitted designs—Nash's
being eventually accepted. The other design cut up the
whole ground into ornamental villas with pleasure
grounds, with a sort of village green or central square,
with a church in the middle, and a site for a market and
barracks. White's views were more like Nash's in some
respects, as he had artificial water and a drive round the
Park. The lease held by the Duke of Portland fell in,
in 1811, and soon after the work of carrying out Nash's
design was begun by James Morgan. The Regent's
Park Canal was included in the same plan, and begun in
1812 and finished in 1820. Its length from Paddington
to Limehouse is 8¾ miles, and the total fall 84 feet.

AUTUMN IN REGENT'S PARK

Although the planting and levelling began in 1812, the buildings rose up slowly. Of the villas in the Park only two were built in 1820, the rent demanded for the ground being extremely high. But two or three years later the whole thing was more or less as it is now, so far as the general outline and buildings are concerned. The cost by May 1826 was £1,533,582, and the estimated probable revenue £36,330. The Prince Regent took the greatest interest in the proceedings, and Nash's design included a site for a palace for him, though even contemporary writers condemned the suggestion, as the situation was damp—" the soil was clay, . . . and the view bad." It was only natural that the Park should henceforth become the Regent's, and not Marylebone 'Park, and the " new street" to connect it with Carlton House be called Regent Street.

It is difficult to judge Regent's Park with an unprejudiced eye. The exaggerated praise it called forth when just completed is only equalled by the unmeasured censure of the next generation. Of the houses which surround it the following are two descriptions. The first, in 1855, calls them " highly-embellished terraces of houses, in which the Doric and Ionic, the Corinthian, and even the Tuscan orders have been employed with ornate effect, aided by architectural sculpture." Fifty years later the same houses are summed up with very different epithets : "Most of the ugly terraces which surround it exhibit all the worst follies of the Grecian architectural mania which disgraced the beginning of this century"! It may not be a style which commends itself to modern taste, but one thing is certain, that having embarked on classical architecture it was best to stick to it and complete the whole. It is as much a bit of history, and as typical of the age, as Elizabethan or Tudor architecture is of theirs, and as such it is best to treat Regent's Park as an interesting example of early nineteenth-century taste.

This ground was country when building was begun, and when one thinks of the streets and crescents that grow up when the country touches the town, and the incongruous ugliness of most of them, there is much to be said for the substantial uniformity of Regent's Park. What can be argued from the surroundings of the other parks? Would Regent's Park have been improved by the erection of rows of houses of the Queen Anne's Mansion type? One cannot help wondering what Stowe would have thought of such a production, when he instances "a remarkable punishment of Pride in high buildings," how a man who built himself a tower in Lime Street, to overlook his neighbours, was very soon "tormented with gouts in his joynts, of his hands and legs"—that he could go no "further than he was led, much lesse was he able to climbe" his tower! What retribution would he have thought sufficiently severe for the perpetrators of Park Row Buildings, New York, with their thirty-two storeys?

Anyhow, Regent's Park was welcomed by the generation who watched it grow. A writer in 1823 says: "When first we saw that Marylebone Fields were enclosed, and that the hedgerow walks which twined through them were gradually being obliterated and the whole district artificially laid out, . . . we underwent a painful feeling or two. . . . A few years, however, have elapsed, and we are not only reconciled to the change alluded to, but rejoice in it. A noble Park is rapidly rising up, and a vast space, close to the metropolis, not only preserved from the encroachment of mean buildings, but laid out with groves, lakes, and villas, . . . while through the place there is a winding road, which commands at every turn some fresh feature of an extensive country prospect." This enthusiast winds up by saying, "We do not envy the apathy of the Englishman who can walk through these splendid piles without feeling his heart swell with national pride." We may smile at such high-sounding language, but, after all, it was an innocent form for national pride to take.

The special feature which the plan of the Park embraced, was the villas, standing in their own pleasure grounds. These were all built in the same Grecian style —most of them designed by Decimus Burton, who was also the architect of Cornwall Terrace, the only one not by Nash. St. Dunstan's Villa, now belonging to Lord Aldenham, and containing his precious library, was his work. It was built by the Marquis of Hertford, and the name is taken from the two giant wooden figures of Gog and Magog, which formerly stood by St. Dunstan's Church in Fleet Street. They had been placed there in 1671, and struck the hours on a large clock (the work of Thomas Harrys), one of the curiosities of the City. It was with reference to them that Cowper's lines on a feeble, uninspired poet were written :—

> "When Labour and when Dullness, club in hand,
> Like the two figures of St. Dunstan's stand,
> Beating alternately, in measured time,
> The clock-work tintinabulum of rhyme,
> Exact and regular the sounds will be,
> But such mere quarter strokes are not for me."

Lord Hertford used to be taken to see them as a child, and had a child's longing to possess the monsters. Unlike most childish dreams, he was able, when the church was rebuilt in 1832, to realise it and to purchase the figures, and remove them to strike the hours in his new villa. St. John's Lodge is another of these detached villas, with a fascinating garden, built by Burton, for Sir Francis Henry Goldsmid; and also in the inner circle there is South Villa, with an observatory, erected in 1837 by Mr. George Bishop, from which various stars and asteroids were discovered by Dawes and Hinde.

The most interesting of the houses in the park is St. Katharine's Lodge, not from any special beauty of its own, but from the sad association of its history. On the east of the road which encircles the Park is St. Katharine's Hospital, built by A. Poynter, a pupil of Nash, in 1827, when the "act of barbarism" of removing the Hospital from the East End was committed. The home of the Hospital, with its church and almshouses, was close to the Tower, and after a peaceful existence of nearly seven hundred years it was completely swept away to make room for more docks. There is nothing to redeem the crude look of uselessness that the new buildings in Regent's Park present. They seem out of place, and as if stranded there by accident. Even thirty years after their removal an official report on the revenues of the hospital shows some signs of repentance. The writers sum up the increased income, then about £11,000 a year, and wonder if in this faraway spot it is being put to the best uses; and the report even goes so far as to suggest its restoration to the populous East End, where the recipients of the charity would spend their lives in the cure of souls, or as nurses and mission-women among the poor. Since then, an improvement has set in as it has become the Central

House for Nurses for the Poor, known as the Jubilee
Nurses, as the funds to provide them were raised by the
women of England as a Jubilee Gift to Queen Victoria.

The Hospital of St. Katharine was founded by Queen
Matilda, " wife to King Stephen, by licence of the Prior
and Convent of the Holy Trinity in London, on whose
ground she founded it. Elianor the Queene, wife to
King Edward the First, a second Foundresse, appointed
to be there, one Master, three Brethren Chaplaines and
three Sisters, ten poore women, and six poore clerkes.
She gave to them the Manor of· Clarton in Wiltshire
and Upchurch in Kent, etc. Queene Philip, wife to
King Edward the Third, 1351, founded a Chauntry
there, and gave to that Hospital tenne pound land by
yeere ; it was of late time [1598] called a free Chappell, a
Colledge and an Hospital for poore sisters. The Quire
which (of late yeares) was not much inferior to that of
Pauls, was dissolved by Doctor Wilson, a late Master
there." Such is Stowe's account of the foundation.

Even in those days the district was becoming crowded,
" pestered with small Tenements," chiefly owing to the
influx from Calais, Hammes, and Guisnes when those
places were lost in Mary's reign. Many, "wanting
Habitation," were allowed a " Place belonging to St.
Katharine's." The curious name, " Hangman's Gains,"
in that locality was said to be derived from a corruption
of two of the places the refugees came from.

In Henry VIII.'s time a Guild or Fraternity was
" founded in the Church of this Hospital of St. Katharine
to the Honour of St. Barbara." Katharine of Aragon
and Henry VIII. and Cardinal Wolsey belonged to it,
and many other " honourable persons." The object was
to secure a home for any " Brother or Sister who fell into

Decay of worldly Goods as by Sekenes or Hurt by the
Warrys, or upon Land or See, or by any other means."
Those belonging to the Fraternity who had paid the full
sum due, namely 10s. 4d., in "money, plate, or any other
honest stufe," were entitled to fourteen pence a week,
house-room and bedding, "and a woman to wash his
clothes and to dresse his mete ; and so to continue Yere
by Yere and Weke by Weke durynge his Lyfe," like a
modern benefit society.   The fine old church contained
many monuments, some of which were transferred to the
new church when the removal took place.   Among them
the effigy of John Holland, Duke of Exeter, and one of
his wives, dating from 1447, reposes under a fine canopy.
The stalls and pulpit of the sixteenth century were also
brought to the new building.   Thus shorn of all its
associations and all its beauty, the foundation remains,
like a flower ruthlessly transplanted too late to take root
and regain its former charm.

The Master's house makes a most delightful residence,
and has always been let.   Mr. Marley, the present tenant,
who has filled the house with works of art, has made a
very charming garden also, more like an Italian than an
English villa garden, as the view reproduced in this
volume testifies.

Three Societies occupy pieces of ground within the
Park.   The most ancient and least well known is the
Toxophilite.   Archery has for many hundred years been
practised by the citizens of London.   The ground chosen
for shooting was chiefly near Islington, Hoxton, and
Shoreditch.   To encourage the use of bows and arrows
Henry VIII. ordered Sir Christopher Morris, Master of
Ordnance, to form the "Fraternitye or Guylde of Saint
George" about 1537, and these archers used to shoot in

Spital Fields. About the time of the Spanish Armada the Honourable Artillery Company was formed, which possessed a company of archers, and for over two hundred years archery was kept alive by this corps, and, following them, by the Finsbury Archers. Just at the time when the corps was abolished Sir Ashton Lever formed the Toxophilite Society in 1781, and the archers of the Honourable Artillery Company became merged in the new Society, which then shot on Blackheath. George IV. belonged to it, and it henceforth became the Royal Toxophilite Society, and settled on ground given to it in Regent's Park in 1834, where it remains, as the lineal descendant of the old historic Guild of Archers. It possesses several interesting relics; a shield given by Queen Mary, and silver cups of the Georgian period, besides a valuable collection of bows and arrows. The hall where the members meet, built when the Society moved to Regent's Park, and added to since, has beneath it some curious cellars with underground passages branching off from them, which it has been suggested may have been part of the outhouses belonging to the Royal Manor House, which stood not far off, on ground now outside the Park. The large iron hooks that were until recently in the cellar walls, seemed suggestive of venison from the Park for the royal table. The ground of the Society is suitably laid out, with a fine sunk lawn for the archery practice. By an arrangement with the Toxophilite Society, "the Skating Club" have their own pavilion, and the lawn is flooded during the winter for their use. There is so much talk about the change of the climate of England, and of the so-called old-fashioned winters, that the record kept by this Skating Club since its foundation in 1830 of the number of

G

skating days in each winter is instructive.   Taking the
periods of ten years during the first decade, 1830–40, there
was an average of 10.2 skating days per winter.   In
1833–34 there were none, in 1837–38 thirty-seven days.
Between 1850–60 the average was only 8.5, while the last
ten years of the century it was 16.8.   It is difficult to
see how any argument could be deduced from such
figures in favour of the excess of cold in the good old
days!   When the freezing of the Thames is quoted to
prove the case, people forget that the Thames has completely changed.   The narrow piers of old London
Bridge no longer get stopped with ice-floes, and the
current is much more rapid now that the whole length is
properly embanked.   In the days when coaches plied from
Westminster to the Temple Stairs as in 1684, or when
people dwelt on the Thames in tents for weeks in 1740,
all the low land was flooded and the stream wider and
more sluggish.   The believers in the hard winters generally maintain the springs were warmer than now, May
Day more like what poets pictured, even allowing the
eleven days later for our equivalent.   But in 1614 there
was snow a foot deep in April, and those who went in
search of flowers on May Day only got snowflakes.   In
1698, on May 8th, there was a deep fall of snow all over
England, and many other instances might be quoted.   So
it seems, though people may grumble now, their ancestors
were no better off.

In the centre of the ground is the Royal Botanical
Society of London, founded in 1839.   At one time the
Society was greatly in fashion, and the membership was
eagerly sought after.   No doubt such will be the case
again, although for some reason the immense advance in
gardening during the last ten years has not met with the

response looked for from this Society, and hence a certain decrease instead of increase in popularity—a phase which can but be transitory. The botanical portions of the grounds illustrative of the natural orders were arranged by James de Carle Sowerby, son of the author of the well-known "English Botany," assisted by Dr. Frederick Farre and others, and the ornamental part of the garden, with the lake, by Marnoch. The designs were severely criticised by Loudon in the first instance, who prophesied failure to the garden, but was well satisfied when the modified plans were announced. Some of the earliest flower shows in the modern sense were held there. And this Society was the pioneer in exhibitions of spring flowers. The first was held in 1862, and was quite a novel departure, although summer and autumn floral shows had been instituted for more than thirty years. These exhibitions and fêtes became very fashionable, and people flocked to them, and numbers joined the Society. It is always difficult to combine two objects, and this is the problem the Botanical Society now has to face. It is almost impossible to keep up the Botanical side and at the same time make a bid for popular public support by turning the grounds partly into a Tea Garden. Now that gardening is more the fashion than it has ever been, it is sad to see this ancient Society taking a back place instead of leading. It is actual horticulture that now engrosses people, the practical cultivation of new and rare plants, the raising and hybridising of florists' varieties. The time for merely wellkept lawns and artificial water and a few masses of bright flowers, which was all the public asked for in the Sixties, has gone by. A thirst for new flowers, for strange combinations of colours, for revivals of long-forgotten plants and curious shrubs, has now taken possession of the large circle of people who profess to be gardeners. Apart from the question whether the present fashion has taken the best direction for the advancement of botany and horticulture, it is evident no society can prosper unless it directs its attention to suit the popular fancy. No doubt this worthy Society will realise this, and emerge triumphant from its present embarrassments.

The third and best known of the societies is the Zoological one. What London child has not spent moments of supreme joy mingled with awe on the back of the forbearing elephant? And there are few grown persons who do not share with them the delight of an hour's stroll through the " Zoo." More than ever, with the improved aviaries and delighful seal ponds, is the Zoo attractive. It was the first of the three Societies to settle in the Park, having been there since 1826. Some of the original buildings were designed by Decimus Burton, who, next to Nash, is the architect most associated with the Park. The Society was the idea of Sir Thomas Raffles, who became the first President in 1825. In three years there were over 12,000 members, and the gardens were thronged by 30,000 visitors. A pass signed by a member was necessary for the admission of every party of people, besides the payment of a shilling each. An abuse of this soon crept in, and people waited at the gates to attach themselves to the parties entering, and well-dressed young ladies begged the kindness of members who were seen approaching the gates. Now only Sunday admittance is dependent on the members. A Guide to Regent's Park in 1829 gives engravings of many of the animals, and shows the summer quarters of the monkeys—most quaint arrangements, like a pigeon

cot on a pole, to which the monkey with chain and ring
was attached, to race up and down at will.

The only alterations of importance after the completion of the Park were the making of the flower
garden, and the filling up of the artificial water to a
uniform depth of 4 feet, after a terrible accident had

STONE VASE IN REGENT'S PARK

occurred in 1867, when the ice broke and forty skaters
lost their lives. The flower-beds are now one of the
most attractive features in the Park, and were originally
designed by Nesfield in 1863. The centre walk continues the line of the "Broad Walk" avenue at its
southern end. In the middle is a fine stone vase supported by griffins, and other stone ornaments in keeping
with the formal style.

The frame-ground in Regent's Park has to be a spacious one, to produce all that is required in the way of spring and summer plants. The fogs are the greatest enemies of the London gardener, and more especially on the heavier soil of Regent's Park. Not even the most hardy of the bedding-out plants will survive the winter, unless in frames. Even wall-flowers and forget-me-nots will perish with a single bad night of fog, unless under glass. Although, on the other hand, it is surprising how some species apparently unsuited to withstand the climate will survive. Among the rock plants growing in a private rock-garden within the Park *Azalia procumbens*, that precarious Alpine, is perfectly at home. Clumps of *Cypripedium spectabele* come up and flower year after year, and *Arnebia echioides*, the prophet flower, by no means easy to grow, seems quite established. But to return to the frame-ground, from whence all the bedding plants emanate. Violas are a special feature in the Park, and one which is much to be commended, as their season of beauty is so protracted. They are all struck in frames, one row of fifty-three lights being devoted to them, in which 23,750 cuttings are put annually. The green-houses are used for storing plants not only for the decoration of the Park but for some fourteen other places outside. The Tower, the Law Courts, Mint, Audit Office, the Mercantile Marine in Poplar, are all supplied from Regent's Park. The Tate Gallery and Hertford House have to be catered for also. Whether the visitors to the Wallace Collection even notice the plants it is impossible to say; they might miss their absence. But the gardeners have to give these few pots considerable care, as they will only stand for a

SPRING IN REGENT'S PARK

very short time inside the building, and after three weeks' visit return to hospital.

Of late years a considerable alteration has been made in the arrangement of the beds in the flower-garden of the Park, chiefly with a view to reducing the bedding and yet obtaining a better effect. Long herbaceous borders have been substituted for one of the rows of formal beds, requiring a constant succession of plants. This has necessitated the removal of some of the flowers shown in the view of this garden taken in the spring. The loss of these is compensated by the new arrangement of beds, separated from the Park by a hedge and flowering shrubs.

Very few of the old trees remain in Regent's Park; what became of them between the time when only a portion were marked for the navy by Cromwell, and the present day, there is no record as yet forthcoming. Two elms near the flower-garden are, however, remarkably fine specimens, as the branches feather on to the ground all round. A *Paulownia tomentosa* is well worthy of notice. It must have been one of the earliest to be planted in this country, and is a large spreading tree. It stands on what is known as the Mound, near Chester Gate. Nineteen years ago it flowered, and in the unusually warm autumn of 1906 it was covered with buds of blossom, all ready to expand, when, alas! the long-delayed frost arrived in October, just too soon for them to come to perfection. Not far from it is a large tree of *Cotoneaster frigida*, which has masses of red berries every year.

The railings of Regent's Park have always been of timber, but it is now threatened to alter this survival of the days when it first changed from Marylebone Farm.

The present timber fence has stood for forty years, so even from an economical point of view iron, which requires painting, could not be recommended. It is to be hoped the old traditional style of fence of this delightful Park may be continued.

To the north of Regent's Park, and only divided from it by a road, lies Primrose Hill. This curious conical hill, 216 feet high, so well known as an open space enjoyed by the public, formerly belonged to Eton College, but became Crown property about the middle of last century, and is now under the Office of Works, who keep it in order, and have done all the planting which has of late years improved this otherwise bare eminence. Some of the guide-books to London refer to the lines of Mother Shipton's prophecy that Primrose Hill "must one day be the centre of London." The passage this is supposed to be based on, is that which used to be said to foretell railways, and now people see in it a foreshadowing of motor cars. At one time also the marriage reference which is in the same poem was applied to Queen Victoria. The lines are these—

> " Carriages without horses shall go,
> And accidents fill the world with woe :
> Primrose Hill in London shall be,
> And in its centre a Bishop's see.
>
> . . . .
>
> The British Olive next shall twine,
> In marriage with the German Vine."

The early editions of the prophecy contain none of these lines except the two last, which are quoted in the 1687 edition, and are there interpreted to refer to the marriage of Elizabeth, daughter of James I., and the

Elector Palatine. The Primrose Hill lines first made
their appearance in 1877! So, although now quite
surrounded by houses, and well within the County of
London, that this would be so in time to come, was not
foretold three hundred years ago.

The delightfully rural name dates from the time of
Queen Elizabeth, and is said to be derived from the
number of primroses which grew there. The earlier
name was Barrow Hill, from supposed ancient burials.
After the mysterious murder of Sir Edmondsbury Godfrey in October 1678, his body was found in a ditch
at the foot of the hill. At one time the superstitious
thought his ghost haunted the place, and a contemporary
medal has this inscription—

" Godfrey walks up hill after he was dead ;
[St.] Denis walks down hill carrying his head."

The fresh air and pleasant view from the top of the
hill, and the cheery sounds of games, have long ago
dispelled all these gloomy memories.

CHAPTER V

GREENWICH PARK

Towered cities please us then,
And the busy hum of men,
Where throngs of knights and barons bold
In weeds of peace high triumphs hold,
With store of ladies, whose bright eyes
Rain influence, and judge the prize
Of wit, or arms, while both contend
To win her grace, whom all commend.
                                    —MILTON.

T would not occur to most people
to reckon Greenwich among the
London Parks. But it is well
within the bounds of the County
of London, and now so easy of
access that it should have no
difficulty in substantiating its
claim to be one of the most beau-
tiful among them. Both for natural features and
historic interest it is one of the most fascinating.

Its Spanish chestnuts are among the distinguishing
characteristics, and although smoke is slowly telling on
them, numbers of these sturdy timber trees are still
in their prime, and it would be hard to find a more
splendid collection in any part of the country. One
of the giants is 20 feet in girth at 3 feet from the
ground, and contains 200 feet of timber.

Those who are the ready champions of the rights of the people to the common lands, and who justly inveigh against all encroachments, must feel bound to admit that, in the case of Greenwich Park, what they would call pilfering in other instances is thoroughly justified. The land which forms the Park was part of Blackheath until Henry VI., in the fifteenth year of his reign, gave his uncle Humphrey, Duke of Gloucester, licence to enclose 200 acres of the wood and heath "to make a park in Greenwich."

The modern history of Greenwich Park may be said to begin in Duke Humphrey's time, but it was a favourite resort long before that. Situated on the high ground above the marshy banks of the river, and near the Watling Street between London and Dover, Greenwich was found suitable for country residence in Roman times. On one of the hills in the Park, with a commanding view over the river, the remains of a Roman villa have been excavated. Over 300 coins were found, dating from 35 B.C. to A.D. 423. Bronzes, pottery, a tesselated pavement, and the remains of painted plaster were discovered, showing that it must have been a villa of "taste and elegance," and there were indications that the final destruction of this charming abode was by fire. A peep into the past might reveal the last of its Roman occupants flying before the barbarian Jute.

Doubtless in its prime there would be a garden near the villa—perhaps a faint imitation of those Roman gardens like Pliny's. There, "in front of the portico," was "a sort of terrace, embellished with various figures and bounded with a box-hedge," which descended "by an easy slope, adorned with the representation of divers animals in box," to a soft lawn. There were shady trees and a splashing fountain, and sunny walks to form " a very pleasing contrast," where the air was "perfumed with roses." The slopes of Greenwich may have presented such a scene in the days when Roman galleys rowed up the Thames.

In another part of the Park, Roman graves have been found, and other burying-places of a later date suggest a very different picture from that of Roman times. These tumuli are very numerous, and although over twenty remain, a much greater number existed, and have been rifled from time to time, or excavated, as in 1784, when some fifty were opened, and braids of human hair, fragments of woollen cloth, and beads were found. These graves suggest the occupation of these heights by the Danes, who were encamped there for some three years about 1011. Wild and lawless must have been the aspect then, and the incident that stands out prominently is the martyrdom of St. Alphege, the Archbishop, slain here by the Danes in 1012.

There was probably some royal residence at Greenwich from the time of Edward I., but it was not until it came to Humphrey, Duke of Gloucester, that the Palace much used in Tudor times was built. This building faced the Thames, and went by the name of "Placentia" or "Plaisance," and round it there was a garden. The royal licence, which gave the Duke leave to enclose a portion of the heath, provided that he might also build "Towers of stone and lime." The tower stood on the hill now crowned by the Observatory, and was pulled down when Charles II. had the Observatory erected from designs by Wren in 1675. The plan included a well 100 feet deep, at the bottom of which the astronomer Flamsteed could lie and observe the heavens.

All through the earlier history of the Park this tower must have been a conspicuous object. During Tudor times Greenwich was much lived in by the Sovereign, and many a gay pageant enlivened the Park. Jousts and tournaments, Christmas games and May Day frolics, were of yearly recurrence in the early days of Henry VIII. The Court moved there regularly to "bring in the May." A picturesque account is given of one of these merry-makings by the Venetian Ambassador and his Secretary. The Ambassador was charmed with the King. "Not only," he writes, is he "very expert in arms and of great valour, and most excellent in personal endowment, but is likewise so gifted and adorned with mental accomplishments of every sort." He joined in the May Day proceedings, which must indeed have presented a brilliant spectacle, with the oaks and hawthorn, and all the wild beauty of Greenwich Park, as a background. Katharine of Aragon, "most excellently attired and very richly, and with her twenty-five damsels mounted on white palfreys, with housings of the same fashion most beautifully embroidered in gold," and followed by "a number of footmen," rode out into the wood, where "they found the King with his guard, all clad in a livery of green with bowers [boughs] in their hands, and about 100 noblemen on horseback, all gorgeously arrayed." "In this wood were certain bowers filled purposely with singing birds, which carrolled most sweetly." Music played, and a banquet under the trees followed, then the procession with the King and Queen together returned to the Palace. The crowds flocking round them the Venetian estimated "to exceed . . . 25,000 persons."

Queen Mary was born at Greenwich, and there she was betrothed to the Dauphin of France. She resided

here much during her short and troublous reign; and
perhaps her fondness for this Palace came from the
association of her early youth, when she was the centre
of attraction. Greenwich cannot always have been
pleasant for the Princess Mary, for here came Anne
Boleyn. From Greenwich she was escorted in state to
London by the Lord Mayor, who was summoned by the
King to fetch her, and from Greenwich she was taken up
the river, her last melancholy journey to the Tower. The
oak under which Henry VIII. is said to have danced
with her is still standing. It is a huge, old, hollow stem,
though quite dead, kept upright by the ivy. The trunk
has a hole 6 feet in diameter, and it is known as Queen
Elizabeth's Oak, as tradition also says she took refresh-
ments inside it. It was fitted with a door, and those who
transgressed the rules of the Park were confined in this
original prison. It was at Greenwich that Queen Elizabeth was born; and to Greenwich Henry brought his
fourth bride, when poor Anne Boleyn's short-lived favour
was at an end, and Jane Seymour dead. The less beautiful
Anne of Cleves, who so signally failed to please the King,
was escorted in state from Calais by thirty gentlemen,
with their servants, "in cotes of black velvet with cheines
of gold about their neckes." On January 3, 1540, the
King rode up from the Palace to meet her on Blackheath
with noblemen, knights, and gentlemen, and citizens, all
in velvet with gold chains. The King rode a horse with
rich trappings of gold damask studded with pearls, a
coat of purple velvet slashed with gold, and a bonnet
decorated with "unvalued gems." Anne came out of
her tent on the Heath to meet him, clad in cloth of gold,
and mounted on a horse with trappings embroidered with
her arms, a lion sable. She rode right through the Park

from the Black Heath to the northern gate and round through the town to the Palace, the guns firing from the Tower in her honour.

It was at Greenwich that the boy king, Edward VI., died, and Mary and Elizabeth were constantly there. Their state barges bearing them to and from the Palace must have been no uncommon sight on the Thames. It was on landing on one of these occasions that the famous episode of Sir Walter Raleigh laying his cloak in the mud for the Queen to tread on, happened. One of the many brilliant scenes in the Park took place after Elizabeth's accession, when the citizens of London, overjoyed, wished to give her a very special greeting. It was on July 2, 1559, that "the City of London entertained the Queen at Greenwich with a muster, each Company sending out a certain number of men-at-arms" (1400 in all), "to her great delight. . . . On the 1st of July they marched out of London in coats of velvet and chaines of gold, with guns, moris pikes, halberds, and flags; and so over London Bridge unto the Duke of Suffolk's Park in Southwark; where they all mustered before the Lord Mayor, and lay abroad in St. George's Fields all that night. The next morning they removed towards Greenwich to the Court there; and thence to Greenwich Park. Here they tarried till eight of the clock; then they marched down into the Lawn, and mustered in arms: all the gunners in shirts of mail. At five of the clock at night the Queen came into the gallery over the Park Gate, with the Ambassadors, Lords, and Ladies, to a great number. The Lord Marquis, Lord Admiral, Lord Dudley, and divers other Lords and Knights, rode to and fro to view them, and to set the two battles in array to skirmish before the Queen:

then came the trumpets to blow on each part, the drums beating, and the flutes playing. There were given three onsets in every battle; the guns discharged on one another, the moris pikes encountered together with great alarm; each ran to their weapons again, and then they fell together as fast as they could, in imitation of close fight. All this while the Queen, with the rest of the Nobles about her, beheld the skirmishings. . . . After all this, Mr. Chamberlain, and divers of the Commons of the City and the Wiflers, came before her Grace, who thanked them heartily, and all the City: whereupon immediately was given the greatest shout as ever was heard, with hurling up of caps. And the Queen shewed herself very merry. After this was a running at tilt. And lastly, all departed home to London."

This fête took place on a Sunday, and the time between the muster and the fight was probably mostly spent in refreshment. The account for the supplies of the "Mete and Drynke" for 1st day of July and Sunday night supper is preserved. They were far from being starved, as, among other items, 9 geese, 14 capons, 8 chickens, 3 quarters and 2 necks of mutton, 4 breasts of veal, beside a sirloin of beef, venison pasties, 8 marrow-bones, fresh sturgeon, 3 gallons of cream, and other delicacies were provided for them. Floral decorations in their honour were not forgotten, and appear in the accounts—"gely flowers and marygolds for iii garlands, 7d.; strawynge herbes, 1/4; bowes for the chemneys, 1d.; flowers for the potts in the wyndowys, 6d."

There is no end to the gay scenes that the Park and even some of the most ancient trees have witnessed. "Goodly banquetting houses" were built of "fir poles decked with birch branches and all manner of flowers both of the field and garden, as roses, gilly flowers, lavender, marigold, and all manner of strewing herbs and rushes " (10th July 1572); and many a brilliant pageant took place under the greenwood tree as well as in the Palace, where Shakespeare acted before the Queen.

Although the days of sumptuous pageantry ended with Elizabeth, much was done for Greenwich by the Stuarts. James I. replaced the wooden fence of the Park by a brick wall, 12 feet high and 2 miles round. At various times sections have been altered or replaced by iron rails, but the greater part of the wall remains as completed between 1619–25.

The "Queen's House," which is the only portion of the older building which still exists, was begun under James I., and completed by Inigo Jones for Queen Henrietta Maria. It was called the House of Delight or the Queen's House, and still bears the latter title. Although the sale does not appear to have been actually completed, Greenwich is among the Royal Parks the Parliament intended to sell. The deer at the time must have been numerous and in good condition, for during the Commonwealth the fear of their being stolen was such, that soldiers were posted in the tower for their preservation. Not any great change, however, took place; the Park remained as it was until completely remodelled by Charles II.

Le Nôtre's name is associated with the changes at Greenwich, as it is with those in St. James's Park, and the style was undoubtedly his; but it is not at all likely that he ever actually came to England, but sent some representative who helped to carry out his ideas. The alterations were under the superintendence of Sir William Boreman, who became Keeper of the Park about that

H

date. In March 1644 John Evelyn made a note in his Diary about planting some trees at his house of Sayes Court, Deptford, and adds, "being the same year that the elms were planted by His Majesty in Greenwich Park." The avenues and all the fine sweet chestnuts were planted about this time, besides coppices and orchards. John Evelyn must have approved of these avenues, as in his "Sylva" he praises the chestnut for "Avenues to our Country-houses; they are a magnificent and royal Ornament." Their nuts were not appreciated in England. "We give that food to our swine," Evelyn continues, "which is amongst the delicacies of Princes in other Countries; . . . doubtless we might propagate their use amongst our common people . . . being a Food so cheap and so lasting."

A series of terraces sloping down from the tower formed part of the design, and their outline can still be traced between the Observatory and the Queen's House, which faces the hill at the foot. Each terrace was 40 yards wide, and on either side Scotch firs were planted 24 feet apart. These trees were brought by General Monk from Scotland in 1664, and until forty years ago many were standing, and the line of the avenue was still traceable; some of the trunks measured 4 feet in diameter at the ground. Smoke tells so much more on all the coniferous tribes than on the deciduous trees, that they have all now perished. The last dead stump had to be felled some ten years ago. The old Palace was much gone to decay when Charles II. began the alterations, so he pulled it down with the exception of the Queen's House, the only part said to be in good repair, and commenced a vast building designed by Wren, one wing of which only was completed in his reign.

Pepys, who always did the right and fashionable thing, of course often went to Greenwich, and mentions many pleasant days there. On one occasion (June 16, 1662) he went "in the afternoon with all the children by water to Greenwich, where I showed them the King's yacht, the house, and the parke, all very pleasant; and to the taverne, and had the musique of the house, and so merrily home again." This excursion having been so successful, he soon after escorted Lady Carteret with great pride, "she being very fine, and her page carrying up her train, she staying a little at my house, and then walked through the garden, and took water, and went first on board the King's pleasure-boat, which pleased her much. Then to Greenwiche Parke; and with much ado she was able to walk up to the top of the hill, and so down again, and took boat . . ." His wife and servants, unencumbered by the fine clothes and the page, had evidently not minded the steep ascent as did this "fine" lady, who, however, was " much pleased with the ramble in every particular of it."

Greenwich Fair was always a great institution, and as a rule it was a riotous and disorderly gathering. Two took place each year, in May and October, and lasted several days. During the seventeenth and following centuries the fairs were notorious, and finally had to be suppressed in the middle of the nineteenth.

When William III. altered the building of Charles II. from a palace to a hospital for seamen in 1694 the Park was kept separate, and the Ranger lived in the "Queen's House." It was not until Princess Sophia held the office in 1816 that the residence was changed to the house which still goes by the name of the Ranger's Lodge, and was lived in by the last Ranger, Lord Wolseley. This

Ranger's House had formerly belonged to Lord Chesterfield, and many of the famous letters to his godson are
dated from there.   No special feature in the garden,
which was thrown open to the public with the Park in
1898, can be attributed to him.   He was not, as Lord
Carnarvon's memoir of him points out, fond of the
country; though he "took some interest in growing
fruit in his garden at Blackheath, he had no love for his
garden like Bacon" or Sir William Temple.   There are
some fine trees in the grounds, especially a copper beech,
with a spread 57 feet in diameter, and a good tulip tree.
Queen Caroline, as Princess of Wales, was Ranger in
1806, and lived in Montague House, since pulled down,
and the "Queen's House" was appropriated to the Royal
Naval School.   At the same time the "Ranger's" was
inhabited by the Duchess of Brunswick, her mother, and
it was on her death that it was purchased by the Crown,
and Princess Sophia, daughter of the Duke of Gloucester,
came to live there as Ranger.   The last royal personage
to stay in the house was the Duke of Connaught, when
studying at Woolwich; and now it serves as refreshment
rooms for the numberless trippers who enjoy Greenwich
Park in the summer.

The most recent changes in the Park have all been
improvements, and now it is beautifully kept.   There is
much that is still wild, and the flora and fauna of the
Park would astonish many.   Among the wild flowers
butcher's broom, spindle, and the parasites on the heather
and the broom, dodder and broom-rape are to be found,
and hart's-tongue, wall rue, polypody and male and lady
ferns.   The list of birds that breed there still is a long
one :—

Barndoor owl.

Spotted fly-catcher.

Missel and the song thrush.

Blackbird.

Hedge sparrow.

Robin.

Sedge and reed warblers.

Black-cap.

White-throat.

The great, blue, and cole tits.

Pied wagtail.

Common bunting.

House sparrow.

Greenfinch.

Linnet.

Bullfinch.

Starling.

Carrion crow.

Jackdaw.

Green woodpecker.

Tree creeper.

Wren.

Nuthatch.

Swallow.

Ring, turtle, and stock doves.

Pigeon.

Moorhen.

Lesser grebe.

The part of the Park fenced off and known as the Wilderness is quiet and undisturbed; there under the big trees, among long grass and bracken, the young fawns are reared every year. They are most confiding and tame—those in the Park too much so; for they are only too ready to eat what is given them, and tragic deaths from a surfeit of orange-peel or such-like delights are the result.

The lake is prettily planted, and red marliac varieties of water-lilies now float on the surface in the summer. The dell, planted with a large collection of flowering shrubs, is well arranged, and many choice varieties, *Solanum crispum*, gum cistus, magnolias, *Buddlea intermedia*, *Indigofera gerardiana floribunda*, and such-like are doing well. The frame-ground is most unostentatious, and it is satisfactory to see how much can be produced. The climate allows of the spring bedding plants and hardy chrysanthemums for autumn being raised out of doors; and the small amount of glass shelters the standard heliotropes, *Streptosolens Jamesoni*, and the like for bedding. Lilies do well in the open; *superbum*, tiger, *thun-*

bergium, Henryii, &c., and pots of *longifolium* flower
strongly after doing duty for three years. There is now
a fair-sized garden, where these plants are displayed,
near the Wilderness, adjoining Blackheath; while the
rest of the Park, with the deer wandering under the
chestnuts, is still left delightfully wild. Under the
shady trees on a summer's day it would still be possible
to dream of Romans and Danes, of pageants and tournaments, and to people the scene with the heroes and
heroines of yore.

CHAPTER VI

MUNICIPAL PARKS

Let cities, kirks, and everie noble towne
Be purified, and decked up and downe.
— ALEXANDER HUME (1557–1609).

ONDON is almost completely sur-
rounded by a chain of parks.
Luckily, as the town grew, the
necessity for fresh air began to be
realised, and before it was too late,
in the thickly-populated districts
north, south, east, and west, any
available open space has been con-
verted into a public garden, or into a more ambitious
park. Would that this laudable spirit had moved people
sooner, and then there might have been a Finsbury Park
nearer Finsbury, and the circle of green patches on the
map might have been more evenly dotted about some of
the intervening parishes. Many of the open spaces are
heaths, or commons, or Lammas Lands, which have
various rights attached to them, and, in consequence,
have been saved from the encroachments which have
threatened them from time to time, and have thus been
preserved, in spite of the growth of the surrounding
districts. Of late years the rights have in many instances
been acquired by public bodies, so as to keep for ever
these priceless boons. It was not until the middle of

last century that the movement in favour of city parks assumed definite form. They were in contemplation before 1840, but none were completed until several years later. Victoria was the first, opened in 1845; Battersea, although begun then, was not ready for planting till 1857; Kennington, Finsbury, and Southwark had followed before 1870, and, since then, every few years new open spaces have been added. They have been purchased by public bodies for the most part, but a large share of the honour of acquiring these grounds is due to private munificence and individual enterprise.

Irrespective of the commons which link them together, the principal parks are the following. Beginning on the extreme north there is Golder's Hill, then to the east of Hampstead lies Waterlow, the next going eastwards is Finsbury, then Clissold and Springfield, and down towards the east Victoria. In South London, between Woolwich and Greenwich, lies Maryon Park; then, west of Greenwich, Deptford and Southwark; then a densely built-over district before Kennington, Vauxhall, and Battersea are reached; while away to the south lie Camberwell, Ruskin, Brockwell, and Dulwich; right away into the country, on the south-east, Avery Hill and Eltham; and back again west, across the river again, in Hammersmith, is Ravenscourt. These parks of varying sizes, and smaller recreation grounds between, make up the actual parks, although some of the commons, with playgrounds, artificial water, and band-stands, can hardly be distinguished from the true park.

The oldest of the parks now under the London County Council—Battersea, Kennington, and Victoria—were for many years under the Office of Works, and on the same footing as the Royal Parks. Government, and no municipal authority, has the credit of their formation. Then came several formed by or transferred to the Metropolitan Board of Works. To all these, already over 2050 acres, the London County Council automatically succeeded. After the Bill reorganising the disposal of the funds of the London Parochial Charities in 1883, a part of their money was allotted to provide open spaces, and they helped to purchase many of the parks—Clissold, Vauxhall, Ravenscourt, Brockwell, and so on. The acquisition of parks has, in many cases, been due to private individuals, who helped to raise the necessary funds, and themselves contributed, and were generally assisted by the local vestries, and, later on, Borough Councils. Miss Octavia Hill, by writing and trying to influence public opinion, made many efforts to secure open spaces. At her instance the Kyrle Society was founded for the general improvement of homes, of disused burial-grounds, and open spaces; and from this developed the Metropolitan Gardens Association, of which the Earl of Meath is Chairman. Immense credit is due to this Society, both for acquiring new sites and beautifying existing ones, and being instrumental in having countless places opened to the public. And to private individuals who have given whole parks, or largely contributed to others, too much gratitude cannot be expressed. Since they came into office, the London County Council has had added some 2300 acres of open spaces and parks to those under its care, which have been purchased, or given in whole or in part, by private individuals or other public bodies. Some of the last acquisitions of the London County Council lie quite outside the county boundary, so are beyond the limit set to this volume. Marble Hill is away at Twickenham, but half the purchase-money of £72,000 was paid by the London County Council, and the entire cost of alteration and maintenance is found by it. The place was bought chiefly to preserve the wooded aspect of the view from Richmond Hill. The Forest of Hainault is also outside the bounds, near Epping. The 805 acres there are partly fields, and in part the remains of the old Forest of "Hyneholt," as it was often written, a section of the Royal Forest which covered a large tract of Essex.

The most natural division, when dealing with these open spaces, is the river, and it is a division which strikes a fairly even balance. Including Royal Parks, which contain some 1266 acres, the northern side can claim the larger area, as, irrespective of squares and churchyards and gardens, there are about 3141 acres of green. The south side has only Greenwich Park of 185 acres of Royal Park, and, exclusive of that, there are quite 2169 acres, as against 1875 of the municipal areas on the northern side, when the Crown land is deducted. Besides these, there are 226 acres maintained by the Borough Councils; so in round numbers London has about 5721 acres of open space. These figures are only rough estimates, and do not include all the smaller recreation grounds or gardens of less than an acre.

These parks scattered around London are enjoyed by hundreds of thousands annually, and yet, to a comparative handful of people who live near Hyde Park, they are as much unexplored country as the regions of Timbuctoo. The bicycling craze of ten years ago suddenly brought Battersea Park into fashion; but the miles of crowded streets, with their rushing trams and top-heavy omnibuses, put a considerable bar between the "West End" and those more distant favoured spots. There is much variety in these parks, both north and south, and the chief difference lies in their origin. When a sub-urban manor-house, standing in its own grounds, with well-timbered park and a garden of some design, has been acquired, a much finer effect is produced than when fields or market-gardens have been bought up and made into a park.

Finsbury Park, for instance, was merely fields, while Waterlow has always been part of a private demesne. It is the same on the south of the river. Brockwell is an old park and garden. Battersea was entirely made. Each park has features which give it an individual character, while there is and must be a certain repetition in describing every one separately.

Many details are of necessity more or less the same in each. The London County Council is responsible for the greater number, and in every case they have thought certain things essential. For instance, the band-stand ; no park, large or small, is considered complete without one. It is hardly necessary to mention each individually, though some are of the ordinary patterns, others more "rustic" in construction (as in Brockwell Park), with branching oak supports and thatched or tiled roofs. Every park, except Waterlow, which is too hilly, furnishes ample area for games. Cricket pitches by the dozen, and space for numerous goal-posts is provided for, in each and all of the larger parks. Gymnasiums, too, are included in the requirements of a fully-equipped park. Swings for the smaller children, bars, ropes, and higher swings for older boys and girls, are supplied. Bathing pools of greater or less dimensions are often added, the one in Victoria Park being especially large and crowded. Then the larger parks have green-houses, and a succession of plants are on view all the year round. The chrysanthemum time is one specially looked forward to in the East End districts. Iron railings and paths, of course, are the inevitable beginnings in the creation of a park, and more or less ambitious gates. It is only in the larger ones, such as Finsbury, Victoria, Dulwich, and Battersea, that carriages are anticipated. Though there is a drive through Brockwell, and the steep hill in Waterlow might be climbed, and the avenue in Ravenscourt is wide enough, it is evidently only foot passengers who are expected, as a rule. Fancy ducks and geese attract the small children on all the ponds, and some parks have enclosures for deer or other animals. Sand gardens, or " seasides " for children to dig in, are also frequently included.

The larger parks are self-contained—that is to say, the bedding out and all the plants necessary for the flower-gardens are reared on the premises. There is a frame-ground with greenhouses attached, where the stock is kept and propagated. Of course, much depends on the soil and locality. In some parks the things will stand the winter much better than in others, where fog and smoke and damp work deadly havoc.

A great deal is now done with simple, hardy flowers, which give just as good an effect as more elaborate and expensive bedding. Roses in the show beds will do well for two or even three years; with a few annuals between they make charming effects. In Finsbury Park, the dark red roses with Canterbury bells, and fuchsias with a ground of alyssum, were effective and simple. In some parks the spring plants will thrive all through the winter. Beds of white Arabis with pink tulips between; forget-me-nots with white tulips; mixed collections of auriculas, that dear old-fashioned "bear's ears," put in about the end of October, make a little show all the winter, and produce a mass of colour in spring. There is still room for improvement in the direction of the planting, but of late years the war waged against the monopoly of calceolarias, geraniums, and blue lobelias has, fortunately, had its effect in a marked degree on the London Parks, municipal as well as royal.

There is apt to be a great uniformity in the selection of plants, more especially among the trees and bushes. The future should always be borne in mind in planting, and alas! that is not always the case. Anything that will grow quickly is often put in, whereas a little patience and a much finer effect would be the result in the end. Privet grows faster than holly, but can the two results be compared? There is a very fine old elm avenue in Ravenscourt; trees which the planter never saw in perfection, but which many generations have since enjoyed. But will the avenue of poplars in Finsbury Park have such a future? After thirty-five years' growth they are considerable trees, but how long will they last? The plane does grow remarkably well, there is no denying, but is it necessary for that reason to exclude almost every other tree? Ash trees thrive surprisingly. Some of the oaks take kindly to London, yet how few are planted. Richard Jefferies, that most delightful of writers on nature, bemoans the lack of English trees in the suburban gardens of London, and the same may be said of the parks to some extent. "Go round the entire circumference of Greater London," he writes, "and find the list ceaselessly repeated. There are acacias, sumachs, cedar deodaras, araucarias, laurels, planes, beds of rhododendrons, and so on." "If, again, search were made in these enclosures for English trees and English shrubs, it would be found that none have been introduced."

It would be even more charming in a London Park than a suburban garden to plant some of the delights of our English country, such as thorns, crab apples, elder, and wild roses, with horse-chestnuts, and hazel. What can be more beautiful than birches at all times of the year? That they grow readily, their well-washed white stems in Hyde Park testify. Birds, too, love the native trees, and some of the songsters, which till lately were plentiful in many parks, might return to build if thus encouraged.

There is much monotony in the laying out of all these parks. The undulating green turf with a wavy line of bushes seems the only recognised form. A narrow strip of herbaceous plants is put between the smutty bushes and well-mown turf, and the official park flower-border is produced. Curving lines of uncertain direction, tortuous paths that carefully avoid the straight line, are all part of the generally received idea of a correct outline. It is always more easy to criticise than to suggest, but surely more variety would be achieved if parks were planted really like wild gardens—the groups of plants more as they might occur in a natural glade or woodland. Then let the herbaceous border be a thing apart—a garden, straight and formal, or curved and round, but not always in bays and promontories jutting into seas of undulating green. A straight line occasionally is a great rest to the eye, but it should begin and end at a definite and tangible point. The small Park in Camberwell has a little avenue of limes running straight across, with a centre where seats can be put and paths diverge at right angles. It is quite small, and yet the Park would be exactly like every other piece of ground, with no particular design, without this.

It gives a point and centre to the meandering paths, and comes as a distinct relief. In Southwark Park an avenue is growing up into fairly large trees. It seems stuck on to the Park—it is not straight, but it is not a definite curve, and it ends somehow by turning towards the entrance at one end and twisting in the direction of the pond at the other. So it remains a shady walk, but not an avenue with any pretension to forming part of a design.

It is not for the formal only this appeal is made, it is for less formality and more real wildness, also a protest against the monotony of the green banks, and bunches of bushes, and meaningless curves, too often the only form of design. The aim in every case must be to have as much variety as possible without incongruity, and to make the utmost use of the ground ; to give the most pleasure at the least expense.

One of the great difficulties must always be the numbers of people who enjoy these parks. The grass suffers to such an extent that portions must be periodically enclosed to recover. Then the children have to be kept at a certain distance from the flowers, or the temptation to gather one over-masters the fear of the park-keeper.

A green walk between trees would be a pleasing change from gravel and asphalt in a less-frequented part of some park, but it would doubtless have to be closed in sections, or there would soon be no turf left; but such an experiment might well be tried. The attempts in Brockwell, Golder's Hill, and Ravenscourt at "old English gardens" are most successful, and a welcome change in the monotony, and one has only to look at the crowded seats to see how much they are appreciated.

The effort to make use of the parks to supplement nature-teaching in the schools is also an advance in the right direction, and one that could be followed up with advantage.

The trials of the climate of London, and the hurtful fogs, must not be forgotten when criticising. They are no new thing, and gardeners for two hundred years have had to contend with the smoke, and wage war against its effects. But the evil has, of course, become greatly intensified during the last fifty years. Fairchild, the author of the "City Gardener," in 1722, regrets that plants will not prosper because of the "Sea Coal." Mirabeau, writing from London in 1784, deplores the fogs in England, and especially "those of London. The prodigious quantity of coal that is consumed, adds to their consistence, prolongs their duration, and eminently contributes to render these vapours more black, and more suffocating—you feel this when rising in the morning. To breathe the fresh morning air is a sort of happiness you cannot enjoy in this immense Capital." Yet in spite of this gloomy picture there are trees now within the London area, which were getting black when Mirabeau wrote. Smuts are by no means solely responsible for trees dying. There are many other contributory causes. The drainage and want of water is often a serious danger, and bad pruning in the case of the younger trees is another. When branches begin to die, it is a very safe and salutary precaution to lop them off, as has lately been done to such a noticeable extent in Kensington Gardens. But the cutting and pruning of trees by those employed by various municipal bodies is often lamentably performed. The branches are not cut off clean, or to a joint, where fresh twigs will soon sprout and fill in and make good the gaps. Often they are cut leaving a piece of wood, which decays back to the young growth, and rots into the sound part of the tree.

Some of the worst enemies of the gardener are the electric power-stations. The trees suffer terribly from the smoke they emit. Even healthy young shrubs and bushes, such as laurels, are destroyed by it. In a very short time they become completely dried up, brown, and shrivelled. In a memorandum on the Electric Power and Supply Bill of 1906, the First Commissioner of Works pointed out these disastrous effects. He says, "The case is not entirely one of the emission or consumption of black or sooty or tarry matters. The other products of combustion, such as sulphurous and sulphuric acid, with solid particles of mineral matter or ash, are very deleterious to vegetation." It appears from the report of Dr. Thorpe, of the Government Laboratory, that the production of sulphuric acid could be "much diminished, if not entirely prevented, by pouring lime-water on the coal before it goes into the furnaces, but from the look of trees in some neighbourhoods this precaution does not appear to be taken." These hindrances are often very disheartening, and the many and serious difficulties that have to be contended with, must never be lost sight of in any review of the parks.

In every case the park is thoroughly appreciated by the inhabitants, and no one can overestimate the health-giving properties of these lungs of the city. It would be vain repetition to point out the fact in each case, or to picture the crowds who enjoy them on

I

Sundays—who walk about, or lounge, or listen to the
bands, or to what appears still more stimulating, to the
impassioned harangue of some would-be reformer or
earnest preacher. The densely-packed audiences, the
gesticulations and heated and declamatory arguments, are
not confined to Hyde Park. Victoria Park gathers just
such assemblies, and every park could make more or
less the same boast. The seats are equally full in each
and all, and the grass as thickly strewn with prostrate
forms. Perambulators are as numerous and children
as conspicuous in the north, south, and eastern parks
as in those of the west.

In looking round the parks it will be well to take
a glance at the smaller ones, then to consider each of
the larger ones more in detail, in every case missing
out some of the obvious appendages which are
characteristic of all.

How pathetic some of these little parks are, and what
a part they play in the lives of those who live in the
dingy streets near. Take, for instance, one with a
high-sounding name, Avondale Park. It is little more
than ten minutes' walk from Shepherd's Bush Station or
Notting Hill Gate. Yet, on inquiry for the most direct
road, nobody can give a satisfactory answer. One man
will say, "I have lived here for years and never heard of
it"; another, "I don't think it can be in this district."
The same would be the result even nearer to it; but ask
for the recreation ground, and any child will tell you.
"Down the first narrow turning and to the right again,
by the pawnbroker at the corner." It is a melancholy
shop, with the plain necessaries of life and tiny babies'
boots for sale on the trays outside the door—what a
volume of wretchedness and poverty those poor things

bespeak. A few yards further, and the iron railings of the "Park" come in view. The happy shrill voices of children resound, the swings are in full motion, the seats well filled, and up and down the asphalt walk, old and young are enjoying themselves. When the band plays the place is packed. "I've calculated as many as nine hundred at one time," says the old guardian, who is proud of the place, "and as for the children, you often can't see the ground for them." Yes, this open space of four and a quarter acres is really appreciated. It is difficult for those in easier circumstances to realise what a difference that little patch of green, those few bright flowers, make to the neighbourhood, or the social effect of the summer evenings, when the band and the pleasant trees offer a counter-attraction to the public-house. For some twelve years this little Park has been enjoyed. Formed by the vestry, and kept up by the Royal Borough of Kensington, it greatly pleases, although it scarce can be called beautiful. The centre is given over to the children, and the boys have ample room, and the girls and infants keep their twenty-four swings in constant motion. A path twists round the irregular plot, and most of the way is bordered by those London-loving plants, the iris, and the usual groups of smutty bushes. Along the front runs a wide asphalt walk, well furnished with seats, a band-stand half way, and a fountain at one end. Some bedding out with gay flowers is the attraction here. A gardener and a boy keep it in order, while for about £20 a year a nurseryman supplies all the necessary bedding-out plants. The old guardian sweeps the scraps of paper up and sees the children are not too riotous at the swings. Thus, for no great expense, widespread pleasure is conferred.

The Embankment Gardens, between Westminster and Blackfriars, are much frequented. At all seasons of the year the seats are crowded, and now, with the statues, bands playing in summer, refreshment buffet, and newspaper kiosk, they look more like a foreign garden than the usual solemn squares of London. During the dinner-hour they are filled with the printers from the many newspaper offices near, and the band was in the first instance paid for by the Press.

They are divided into three sections, and measure ten acres in all, not including the garden beyond the Victoria Tower. The peace has been utterly destroyed by the din of trams, which are for ever passing and re-passing, and it is much to be feared that the trees next the river, which were growing so well, will not withstand the ill-treatment they have received—the cutting of roots and depriving them of moisture. The Gardens are entirely on the ground made up when the Embankment was formed, between 1864 and 1870.

The Gardens were opened in 1870, but many improvements have since been made in the design, and various statues put up to famous men. One is to John Stuart Mill, and at the Westminster end, one of William Tyndall, the translator of the New Testament and Pentateuch, to which translation is due much of the beautiful language of the Authorised Version of the Bible.

Of the old gardens and entrances to the great houses which stretched the whole length of the river bank, from Westminster and Whitehall to the City, only one trace remains. It is the Water Gate of York House. The low level on which it stands, below the terrace end of Buckingham Street, shows to what point the river rose. York House was so called as it was the town house of the Archbishops of York, but none of them ever lived there except Heath, in Queen Mary's time, who was the first to possess it. It was let, as a rule, to the Keepers of the Great Seal, and Bacon lived there. George Villiers, Duke of Buckingham, pulled down most of the old house, and commenced rebuilding. Nothing now remains but the Water Gate, supposed to be by Inigo Jones, although the design is also attributed to Nicholas Stone, who built it. The house and gardens were sold and divided in 1672. Buckingham Street and the streets adjoining are built on the site, and all that is left is the fine old gateway, with most modern-looking gardens between it and the river, which once flowed up to its arches.

Another Embankment recreation ground is the Island Garden, Poplar, and it is one that is also much appreciated. It was made on some ground not required for ship-building or docks on the river front of the Isle of Dogs, and opened to the public in 1895. The idea of making a garden of it had for some few years been in contemplation, and as soon as the necessary funds were found, this space, somewhat less than three acres, was saved from being built over, and a wide walk of about 700 feet made along the river embankment. The view from the seats, with which it is plentifully supplied, over towards Greenwich Hospital and Park makes it a really charming promenade. The quaint name of this part of London is said to be derived from the fact that the kennels of the sporting dogs of the royal residents of Greenwich Palace were kept there, " which usually making a great noise, the seamen and others thereupon called the place the Isle of Dogs." This seems the most plausible of the various definitions of the name of this peninsula—for it is only

an island by means of the dock canal, made in 1800.
A quotation from a play of Middleton and Dekker,
in 1611, shows that then, at any rate, it was associated
with actual dogs.

"*Moll Cutpurse:* O Sir, he hath been brought
up in the Isle of Dogs, and can both fawn like a
spaniel and bite like a mastiff, as he finds occasion."

The ground in those days and until much later
times was a fertile marsh, subject to frequent inundations, but affording very rich pasture. Breaches in the
embankment occurred at intervals until a solid pile and
brick wall was made in the last century, above which
the "Island Gardens" were laid.

Further along the north bank of the river there is
another and a larger garden, kept up by the London
County Council, although it is in East Ham and not
within the County of London. This was made on
the site of the North Woolwich Tea Gardens, which
enjoyed a kind of popularity for some fifty years.
Having been started in 1851, they kept up their reputation for "Baby Shows," "Beard Shows," and such-
like attractions, until the ground became too valuable
for building, and too heavily rated for them to exist,
and, but for timely interference, this open space would
have been converted into wharves.

The story of the Bethnal Green Gardens is very
different. Although it was only in 1891 that the
present arrangements with regard to keeping up the
Gardens were established, the 15½ acres of which they
form part has a long history. As far back as 1667
the land was purchased by a group of residents, who
collectively suscribed £200, and by a trust-deed dated
1690 conveyed the land to trustees, to be administered

for the benefit of the poor. It had been purchased and enclosed, the deed specified, "for the prevention of any new building thereon." Of this ground 9 acres form the present Garden; on part of the remainder St. John's Church was built, and in 1872 the Bethnal Green Museum, an offshoot from South Kensington, was opened on another section. The most exhaustive work on Municipal Parks says that when the land "came into the possession of the London County Council" it "consisted of orchard, paddock, kitchen garden, and pleasure grounds, all in a rough and neglected condition." Under the levelling hand of the London County Council it has been made to look exactly like every other public garden, with "ornamental wrought-iron enclosing fences, broad walks, shrubberies," and so on, at a cost of over £5000, and was opened in 1895. There is no trace of its former condition, nothing to point to its antiquity or any difference in its appearance from the most modern acquisition. Perhaps after all it is as well, for among the thousands of that poor and crowded district that use and enjoy it, there is not one to whom a passing thought of the old weavers who were settled there when the land was given, or to whom the legend of pretty Bessee the Blind Beggar's daughter of Bethnal Green would occur. Though the design is prosaic, the gardens are made cheerful and gay, and if they add a gleam of brightness to the lives of toil of those living near them, they must be said to fulfil their purpose.

Victoria Park

Victoria Park was the first of the modern Parks to be laid out, and it is the largest. When the advantage

of an East End Park was admitted, the work of
forming one was carried out by the Commissioners of
Woods and Forests. An Act passed in 1840 enabled
them to sell York House to the Duke of Sutherland
(hence it became Stafford House), for £72,000, and
to purchase about 290 acres of land in the East End
in the parishes of Hackney, Bethnal Green, and Bow.
Part of this was reserved for building improved dwellings, and 193 acres formed Victoria Park, the laying
out of which began in 1842. Thirty years later, when
some of the land adjoining was about to be built on,
the Metropolitan Board of Works bought some 24 acres
to add to the Park, the whole of which, including
the new part, was under the Office of Works. Other
additions have been made from time to time, chiefly
with a view to opening entrances to the Park, so as
to make it as easy of access as possible from the crowded
districts in the direction of Limehouse and the docks,
and round Mile End Road.

The ground which the Park covers was chiefly
brick-fields and market-gardens, and Bishop's Hall
Farm. The latter place is the only part with any
historical association. The farm was in the manor of
Stepney, which was held by the Bishops of London,
and Bishop's or Bonner's Hall was the Manor House.
Many of the Bishops of London resided here in early
days. Stowe, in 1598, referring to Bishop Richard de
Gravesend in 1280, writes: " It appeareth by the
Charter [of free] warren granted to this Bishop, that (in
his time) there were two Woods in the Parish of Stebun-
heth [Stepney], pertaining to the said Bishop: I have
(since I kept house for my selfe) knowne the one of
them by *Bishops Hall*, but now they are both made

plaine of wood, and not to be discerned from other grounds." These woods were on the ground covered by the Park. Stowe notices in his short accounts of the Bishops of London that Ralph Stratford, who was Bishop from 1339 to 1354, "deceased at Stebunhith." The name Bonner's Hall somehow became attached to the Manor House. The same chronicler also records that Bishop Ridley gave the manors of Stepney and Hackney to the King in the fourth year of Edward VI., who granted them to Lord Wentworth. Bonner, therefore, would be the last Bishop who could have resided there. The old Manor House was not destroyed till 1800, when part of the material was taken to build a farm-house, which was cleared away when the Park was formed.

The first laying out of the Park does not seem to have been altogether satisfactory. A writer in 1851 criticises it very severely. The roads and paths, he says, were so badly laid as to require almost reconstruction. The "banks of the lake must be reduced to something like shape to resist the wash of the water," and the remodelling of the plantations will be "a work of time." Just then Mr. Gibson assumed the charge of the Park, and even this captious critic seems to have been well satisfied that he had "begun in real earnest" to carry out the necessary improvements. Modern gardeners might not applaud all his planting quite so enthusiastically as his contemporaries. For instance, the rage for araucarias—monkey puzzles—has somewhat subsided, though the planting of a number met with great praise in the Fifties. Most of the Park was planted with discrimination. In a line with the canal which forms one boundary, an avenue was put, now a charming

shady road with well-grown trees.   The artificial water
with fancy ducks, in which is a wooded island with
a Chinese pagoda, is a great delight for boating.   The
bathing-lake has still greater attraction, and thousands

PAGODA ON THE ISLAND, VICTORIA PARK

bathe there daily all through the summer months.
It is said, as many as 25,000 have been counted on a
summer's morning.   Bedding out was at its height
when Victoria Park was laid out, so the flower-garden
included some elaborate scroll designs which were
suited to the style of carpet-bedding then in vogue.
Now, though less stiff, the formal bedding is well done,

and attracts great attention. Those in the East End have just as keen an appreciation as the frequenters of Hyde Park, of the display of flowers. The green-house in winter is much enjoyed, and a succession of bright flowers is kept there during the dark months of the year. The children's sand garden is also a delight.

In spite of its situation in a densely-populated district, the feathered tribes have not quite deserted the Park. The moor-hen builds by the lake and the ringdove nests in the trees. Though the greenfinch and the wren have vanished, some songsters still gladden the world. Blackbirds, thrushes, and chaffinches are by no means uncommon. Some of these latter get caught, and take part in the popular amusement of singing-matches. Many men in the district keep chaffinches in cages, and bring them to the Park on a Sunday morning that they may practise their notes in chorus with their wild associates, and so beat the caged bird of some rival. Sometimes the temptation is too great, and the wild birds are kidnapped to join the competition.

FINSBURY PARK

Finsbury is second in size, and second in date of construction, of the Parks of North London. It is far from Finsbury, being really in Hornsey, but as the idea, first expressed about 1850, was to make a Park for the borough of Finsbury, the name was retained although the land acquired some years later was somewhat remote.

The movement was first set on foot when building began to destroy all the open spaces near Finsbury Fields. Some of these, like Spa Fields, had been popular places of resort as Tea Gardens, but were being rapidly covered with houses, and separating Finsbury altogether from the country.   Many delays, owing to changes of Government, occurred before the necessary legislation was accomplished. When the Metropolitan Board of Works came into being, it took up the scheme, and it was finally under its auspices that the land was purchased, and the Park, 115 acres in extent, was opened in 1869.

On the highest point of the ground there is a lake, which was in existence before it became a public park.   Near there stood Hornsey Wood House, a Tea Garden of some reputation in the eighteenth century. About the year 1800 the old house was pulled down, and the new proprietor built another tavern, and converted part of the remains of Hornsey Wood into an artificial lake for boating and angling.   This second house existed until it was pulled down in 1866, when the Park was in progress.   Hornsey Wood was part of the forest which bounded London on the north, and the site of the Park was in the manor of Brownswood, which was held by the See of London.

Accounts of various incidents which are connected with this spot are given in histories of Hornsey.   The most picturesque is that in which the ill-fated little King Edward V. is the central figure, overshadowed by his perfidious uncle.   "The King on his way to London [from Ludlow] was on the fourth of May met at Hornsey Park (now [1756] Highgate) by Edmund Shaw, the Mayor, accompanied by the Aldermen, Sheriffs and five hundred Citizens on Horseback, richly accoutered in purple Gowns; whence they conducted him to the City; where he was received by the Citizens with a joy inexpressible. . . . In this solemn Cavalcade, the Duke of Gloucester's Deportment was very remarkable; for riding before the King, uncovered, he frequently called to the Citizens, with an audible voice, to behold their Prince and Sovereign." What a scene must the site of Finsbury Park have presented that May morning. The Londoners, incensed at Gloucester's having taken possession of the young King, no doubt meet him with distrust and anger, and while the procession moves on towards the City he allays their suspicions, acting a part to deceive them.

The trees in Finsbury are beginning to grow up, and the Park is losing the new, bare look which made it unattractive in its early years. Poplars (fast-growing trees) have been largely used. That is very well for a beginning, but others of a slower growth, but making finer timber, are the trees for the future. There is nothing very special to notice in the general laying out of the grounds, as beyond the avenue of black poplars and the lake, there are no striking features. The view from the high ground, towards Epping, adds to the attractions of this useful open space but not very interesting Park. One of the most pleasing corners is the rock garden, not far from the lake. The plants seem well established and very much at home. The greenhouses, too, are well kept up, and in the gloomy seasons of the year especially are much frequented.

CLISSOLD PARK

Clissold, or Stoke Newington Park, is one of the parks which has the advantage of having been the grounds of a private house, and enjoys all the benefits of a well-planted suburban demesne. The old trees at once give it a certain *cachet* that even County Council railings, notice-boards, and bird-cages cannot destroy.

It has the additional charm of the New River passing through the heart of it, and, furthermore, the ground is undulating.

One of the approaches to the Park still has a semi-rural aspect and associations attached to it. This is Queen Elizabeth's Walk, with a row of fine elm trees, under which the Queen may have passed as a girl while staying in seclusion at the manor-house, then in the possession of the Dudley family, relations to the Earl of Leicester. Stoke Newington, until lately, was not so overrun with small houses as most of the suburbs. In 1855 it was described as " one of the few rural villages in the immediate environs [of London]. Though, as the crow flies, but three miles from the General Post Office, it is still rich in parks, gardens, and old trees." The last fifty years have quite transformed its appearance. " Green Lanes," which skirts the west of the Park, though with such a rural-sounding name, is a busy thoroughfare, with rushing trams; and, but for Clissold Park and Abney Park Cemetery, but little of its former attractions would remain. The Cemetery is on the grounds of the old Manor House, where Sir Thomas Abney lived, and " the late excellent Dr. Isaac Watts was treated for thirty-six years with all the kindness that friendship could prompt, and all the attention that respect could dictate." The manor was sold by direction of Sir Thomas's daughter's will, and the proceeds devoted to charitable purposes. The old church, with its thin spire, and the new large, handsome Gothic church, built to meet the needs of the growing population, stand close together at one corner of the Park, at the end of Queen Elizabeth's Walk, and on all sides the towers among the trees form pretty and con-spicuous objects. The house in the Park, for the most

part disused, stands above the bend of the New River,
which makes a loop through the grounds. It is a white
Georgian house with columns, and looks well with
wide steps and slope to the water's edge, now alas!

STOKE NEWINGTON CHURCH FROM CLISSOLD PARK

disfigured by high iron railings. The place belonged to
the Crawshay family, by whom it was sold. The
daughter of one of the owners had a romantic attachment to a curate, the Rev. Augustus Clissold, but the
father would not allow the marriage, and kept his

daughter more or less a prisoner. After her father's death, however, she married her lover, and succeeded to the estate, and changed its name from Crawshay Farm to Clissold Place. This title has stuck to it, although it reverted to the Crawshays, and in 1886 was sold by them.

The Park measures 53 acres. There is a small enclosure with fallow deer and guinea-pigs, some artificial water, and wide green spaces for games; but the special beauty of the Park consists in the canal-like New River, with walks beside it, and in places foliage arching over it, and the fine large specimen trees round the house. There are some good cedars, deciduous cypress, ilex, thorns, and laburnums; a good specimen of one of the American varieties of oak, *Quercus palustris*; also acacias and chestnuts—all looking quite healthy.

Springfield Park

Not very far from Clissold lies Springfield Park, in Upper Clapton, opened to the public in 1905. It also has the advantage of being made out of well laid out private grounds. The area, 32½ acres, embraced three residences, two of which have been pulled down, while the third, Springfield House, which gives its name to the Park, has been retained, and serves as refreshment rooms. The view from the front of the house over Walthamstow Marshes is very extensive. The ground slopes steeply to the river Lea, and beyond on the plain, like a lake, the reservoirs of the "East London Works," now part of the Metropolitan Water Board, make a striking picture. Springfield House was, until lately, one of those pleasant old-fashioned residences of which there were many in this neighbourhood, standing in well-planted gardens overlooking the marshes and fertile flats below. These delightful houses are becoming more rare every year, and it is fortunate that the grounds of one of the most attractive should have been preserved as a public park. The place was well cared for in old days, as the good specimen trees testify. A flourishing purple beech is growing up, also a sweet chestnut and several birches. A very old black mulberry still survives, although showing signs of age. There are other nice timber trees on the hillside, and among the shrubs an *Arbutus unedo*, the strawberry tree, is one of the most unusual. This Park, though small, is quite unlike any other, and has much to recommend it to the general public, while in the more immediate neighbourhood it is greatly appreciated.

WATERLOW PARK

Undoubtedly the most beautiful of all the parks is Waterlow, the munificent gift of Sir Sydney Waterlow. Its situation near Highgate, above all City smoke; its steep slopes and fine trees; its old garden and historic associations, combine to give it a character and a charm of its own. It is small in comparison with such parks as Victoria, Battersea, or Finsbury, being only 29 acres, but it has a fascination quite out of proportion to its size. There are few pleasanter spots on a summer's day, and at any season of the year it would well repay a visit. It is especially attractive when the great city with its domes and towers is seen clearly at the foot of the hill. London from a distance never looks hard and sharp and clear, like some foreign towns. The buildings do not stand up in definite outline like the churches of Paris looked down upon from the Eiffel Tower: the

K

soft curtain of smoke, the mysterious blue light, a gentle reminder of orange and black fog, shrouds and beautifies everything it touches. On a June day, when the grass is vivid and the trees a bright pale green, Waterlow Park is at its best. The dome of St. Paul's, the countless towers of Wren's city churches, the pinnacles of the Law Courts, the wonderful Tower Bridge, dwarfing the old Norman White Tower, all appear in softened beauty behind the fresh verdure, through well-contrived peeps and gaps in the trees.

Most of the ground is too steep for the cricket and football to which the greater part of other parks are given over. Only lawn tennis and bowls can be provided for, on the green lawns at the top of the Park. A delightful old pond, with steep banks overshadowed by limes and chestnuts, has a feeling of the real country about it. The concrete edges, the little patches of aquatic plants and neat turf, are missing. The banks show signs of last year's leaves, fallen sticks, and blackened chestnuts, and any green near it, is only natural wild plants that enjoy shade and moisture. It is the sort of place a water-hen would feel at home in, and not expect to meet intruding Mandarin ducks or Canadian geese. Let us hope this quiet spot may long remain untouched. There are two newer lakes lower down, laid out in approved County Council style, trim and neat, with water-fowl, water-lilies, and judicious planting round the banks of weeping willows and rhododendron clumps. Probably many visitors find them more attractive than the upper pool. There is no fault to find with them, and they are perhaps more suited to a public park, but they are devoid of the poetry which raises the other out of the commonplace. As the slopes towards the lower lakes are the playground of multitudes of babies, it is necessary to protect them from the water's edge by substantial railings, but most of the Park is singularly free from these unsightly but often necessary safeguards. The trees all through the grounds are unusually fine. Four hickories are particularly worthy of note. They are indeed grand and graceful trees, and it is astonishing they should be so little planted. These are noble specimens, and look extremely healthy.

The most characteristic feature in the Park is the house it contains and the garden immediately round it. This was built for Lauderdale, the " L " in the Cabal of Charles II., probably about 1660. When this unattractive character was not living there himself, he not unfrequently lent it to Nell Gwynn. The ground floor of the house is open to the public as refreshment rooms, and one empty parlour with seats has much good old carving, of the date of the house, over the mantelpiece, also in a recess which encloses a marble bath known as " Nell Gwynn's bath." It is said to have been from a window in Lauderdale House that she held out her son when Charles was walking below, threatening to let him drop if the King did not promise to confer some title upon him. In response Charles exclaimed, " Save the Earl of Burford," which title (and later, that of Duke of St. Albans) was formally conferred upon him.

The terrace along which the King was walking is still there. A little inscription has been inserted on a sun-dial near the wall, to record the fact that the dial-plate is level with the top of St. Paul's Cathedral. A flight of steps leads to a lower terrace. This is planted in a formal design consisting of three circles, the centre one having a fountain. Two more flights of steps descend, in a line from the fountain, to a broad walk bordered with flowers leading to one of the entrances to the Park. At right angles to the other steps a walk leads from the fountain to another part of the garden, which is planted with old fruit-trees on the grassy slope. It is at the foot of these steps that the water-colour sketch is taken. The "eagles with wings expanded" are the supporters of the Lauderdale arms. The whole garden is delightful, and so much in keeping with the house that it is easy to picture the much-disliked Lauderdale, the genial King, and fascinating "Nell," living and moving on its terraces. Pepys gives a glimpse of one of these characters at home. He drove up alone with Lord Brouncker, in a coach and six. No doubt the hill made the six very necessary, as in another place Pepys talks of the bad road to Highgate. They joined Lord Lauderdale "and his lady, and some Scotch people," at supper. Scotch airs were played by one of the servants on the violin; "the best of their country, as they seemed to esteem them, by their praising and admiring them: but, Lord! the strangest ayre that ever I heard in my life, and all of one cast. But strange to hear my Lord Lauderdale say himself that he had rather hear a cat mew, than the best musique in the world; and the better the musique, the more sick it makes him; and that of all instruments, he hates the lute most, and next to that the baggpipe." These sentiments may not prove that Lauderdale was "a man of mighty good reason and judgement," as Lord Brouncker assured Pepys when he said he thought it "odd company," but at least it shows him honest! How many people who sit patiently through a performance of the "Ring" would have as much courage of their opinions?

WATERLOW PARK

Within the grounds of the present Park, near Lauderdale House, stood a small cottage in which Andrew
Marvel lived, which was only pulled down in 1869. It
was considered unsafe, and no National Trust Society
was then in existence to make efforts for its preservation.
In a "History of Highgate" in 1842 the connection
between the place and this curious personage, political
writer, poet, Member of Parliament, and friend of Milton
is barely commented on. "Andrew Marvel, a writer
of the seventeenth century, resided on the Bank at Highgate in the cottage now occupied by Mrs. Walker." The
reader of these lines is penetrated with a feeling that he
ought to know all about Mrs. Walker, rather than the
obscure writer!

The kitchen-garden is large, with charming herbaceous borders, and a long row of glass-houses and vineries,
and the grapes produced have hitherto been given to
hospitals. Let us hope that the same complaint will not
arise here as in another Park, where out-door fruit was
distributed, and caused such jealousies that the practice
was discontinued.

With such a high standard set by the existing gardens,
it is curious that the new bedding should be as much out
of harmony as possible. The beds which call forth this
remark are those round the band-stand. The shape of
them it is impossible to describe, for they are of uncertain
form and indistinct meaning. The flowers are in bold
groups, and yet they look thoroughly out of place.

Wandering one summer's day near the statue, erected
to Sir Sydney Waterlow, the writer overheard some girls,
who looked like shop-girls out for a holiday, discussing
who it was. The most enterprising went up and read
the inscription. "To Sir Sydney H. Waterlow, Bart.,

donor of the Park 1889, Lord Mayor of London 1872–73. Erected by public subscription 1900." "Why, it's to some chap that was once Lord Mayor," was the remark to her friend, following a close scrutiny of this bald inscription. The impulse to explain the meaning of the word "donor" was irresistible; it was evidently quite Greek to these two Cockney young ladies. On learning the meaning they were very ready to join in a tribute of gratitude to the giver of such a princely present. Surely a few words expressing such a feeling would have been appropriate on the statue so rightly erected in memory of the gift! Profound feelings of thanks to the giver must indeed be experienced by every one who has the privilege of enjoying this lovely Park, one of the most charming spots within easy reach of the heart of the City.

GOLDER'S HILL PARK

Golder's Hill Park joins the western end of Hampstead Heath, but its park-like appearance and house and garden are quite a contrast to the wilder scenery of the Heath, although Golder's Hill seems more in the country than Hampstead, as the houses near are so well hidden from it. The mansion has a modern exterior, although parts of it are very old, and the fine trees in the grounds show that it has been a pleasant residence for some hundreds of years. The estate of 36 acres was bought in 1898 from the executors of Sir Spencer Wells, the money in the first instance being advanced by three public-spirited gentlemen, anxious to save the charming spot from the hands of the builder. The view from the terrace of the house, which now serves as a refreshment room, is very pretty, with a gently sloping lawn in front, park-like meadows, and fine trees beyond the dividing sunk fence, and distant peeps of the country towards Harrow. The approach from the Finchley Road is by an avenue of chestnuts, and a flat paddock on one side is a hockey and cricket-ground for ladies. There are some really fine oaks, good beeches, ash, sycamore, Spanish chestnuts, and Scotch firs; but the most remarkable tree is a very fine tulip, which flowers profusely nearly every year. At the bottom of the Park an undisturbed pond, with reedy margin, is much frequented by moor-hens. The valley above is railed off for some red deer, peacocks, and an emu, while three storks are to be seen prancing about under the oak trees in the open Park. The most attractive corner is the kitchen-garden, which, like the one in Brockwell, has been turned into an extremely pretty flower-garden. On one side is a range of hothouses, where plants are produced for bedding out, and a good supply of fruit is raised and sold to the refreshment-room contractor on the spot. Two sides have old red walls covered with pear.trees, which produce but little fruit, and the fourth has a good holly hedge. The vines from one of the vineries have been planted out, and they cover a large rustic shelter, and have picturesque though not edible bunches of grapes every year. The way the planting of roses, herbaceous and rock plants, and spring bulbs is arranged is very good; but the same misleadingly-worded notice with regard to the plants of Shakespeare is placed here as in the Brockwell "old English garden."[1] There is a nice old quince and other fruit-tree standards in this really charming garden. In another part of the grounds there is an orchard, not "improved" in any

[1] See p. 171.

way, but left as it might be in Herefordshire, with grass
and wild flowers under the trees, which bear bushels of
ruddy apples every year.

Part of the Park is actually outside London, but it is
all kept up by the London County Council. The parish
boundary of Hampstead and Hendon, which is also the
limit of the County of London, is seen in the middle
among the oak trees.

RAVENSCOURT PARK

Ravenscourt is another of those parks the nucleus of
which was an old Manor House, hence the existence
of fine old trees, which at once lift from it the crudeness
which is invariably associated with a brand-new Municipal
Park. A bird's-eye view of the ground is familiar to
many who pass over the viaduct in the London and
South-Western trains. These arches intersect one end of
the Park, and cut across the beginning of the fine old
elm avenue, one of its most beautiful features. A bright
piece of garden, typical of every London Park, with raised
borders in bays and promontories, jutting into grass and
backed by bushes, lies to the south of the viaduct.
Where two paths diverge there is a pleasing variation to
the usual type—a sun-dial—erected by Sir William Bull
to "a sunny memory." The arches have been utilised
so as to compensate for the intrusion of the railway.
Asphalted underneath, they form shelters in wet weather
—one is given over to an aviary, two to bars for the
elder children to climb on, and one is fitted with swings
for the babies. This arch is by far the most popular,
and it requires all the vigilance of the park-keeper to see
that only the really small children use the swings, or
the bigger girls would monopolise them. Perhaps the

indulgent and fatherly London County Council will provide swings for the elders, too, some day, and so remove the small jealousies.

To the west of the long avenue lies the orchard. A stretch of grass, devoted to tennis-courts and bowling-greens, separates the pear trees from the walk. These pears and the solitary apple tree are delightful in spring, and a temptation in autumn. Round the house, which is not by any means as picturesque as the date of its building (about 1649) would lead one to expect, are some good trees—planes that are really old, with massive stems, horse-chestnuts and limes, acacias that have seen their best days, cedars suffering from age and smoke, and a good catalpa. The Manor House which preceded the present building was of ancient origin. In early times it was known as the Manor House of Paddenswick, or Pallenswick, under the Manor of Fulham, and was the residence of Alice Perrers, the favourite of Edward III. It was seized in 1378, when she was banished by Richard II.; but after the reversion of her sentence, she returned to England as the wife of Lord Windsor, and the King, in 1380, granted the manor to him. It is not heard of again till Elizabeth's time, when it belonged to the Payne family, and was sold by them in 1631 to Sir Richard Gurney, the Royalist Lord Mayor, who perished in the Tower. After his death it was bought by Maximilian Bard, who probably pulled down the old house and built the present one, which is now used as the Hammersmith Public Library. In the eighteenth century the name was changed from Paddenswick (a title preserved by a road of that name running near the Park) to Ravenscourt, an enduring recollection of the device of a black raven, the arms of Thomas Corbett,

Secretary to the Admiralty, who owned the place for a few short years. Nearly every vestige of the surroundings of the old manor was obliterated and improved away by Humphrey Repton, the celebrated landscape gardener. He filled up most of the old moat, except a small piece, which was transformed into a lake, more in harmony with the landscape school to which he belonged. This piece of water is a pretty feature in the Park, and an attempt has been made to recall the older style, by introducing a little formal garden in an angle of the enclosing wall of the Park. The square has been completed with two hedges, one of them of holly, and good iron gates afford an entrance. The " old English garden," from which dogs and young children, unless under proper supervision, are excluded, is laid out in good taste—a simple, suitable design, with appropriate masses of roses and herbaceous plants, arches with climbers, and an abundance of seats. It has the same misleading notice with regard to Shakespearian plants, as in Golder's Hill and Brockwell, one of the South London Parks, which must now be looked at.

CHAPTER VII

MUNICIPAL PARKS IN SOUTH LONDON

No fresh'ning breeze—no trellised bower,
No bee to chase from flower to flower ;
'Tis dimly close—in city pent—
But the hearts within it are well content.
— ELIZA COOK.

F the South London Parks Battersea is the largest and most westerly, and the best known to people outside its own district. Battersea is entirely new, and has no history as a Park, for before the middle of last century the greater part was nothing but a dismal marsh. The ground had to be raised and entirely made before the planting of it as a park could begin at all. The site was low-lying fields with reeds and swamps near the water, and market-gardens famous for the asparagus, sold as "Battersea bundles," growing around it. In the eighteenth century three windmills were conspicuous from the river. One ground corn, another the colours, and the third served to grind the white lead for the potteries. This was during the time when Battersea enamel was at its height, and snuff-boxes were being turned out in quantities. On the banks of the river stood a tavern and Tea Garden, known as the Red House for many generations. It was much resorted to, but latterly its reputation was none of the best. Games of all kinds took place in its gardens, and pigeon-shooting was one of the greatest attractions there, during the first half of the nineteenth century. Although for long, crowds enjoyed harmless amusements there—"flounder breakfasts," and an annual "sucking-pig dinner," and such-like—towards the end of the time of its existence, it became the centre of such noisy and riotous merrymakings that the grounds of the Red House became notorious. The Sunday fairs, with the attendant evils of races, gambling, and drinking, were crowded, and thousands of the less reputable sections of the community landed every Sunday at the Red House to join in these revellings. It was chiefly with a view to doing away with this state of affairs, that the scheme was set on foot, for absorbing the grounds of the Red House, and other less famous taverns and gardens that had sprung up round it, and forming a Park.

Battersea, or "Patricesy," as it is written in Domesday, was a manor belonging to the Abbey of Westminster until the Dissolution of the Monasteries. The name is most probably derived from the fact that it was lands of St. Peter's Abbey "by the water." Later on it came into the St. John family, and Henry St. John, Lord Bolingbroke, was born and died in Battersea. After his death it was purchased by Earl Spencer, in whose family it remains. Part of the fields were Lammas Lands, for which the parish was duly compensated. The gloomy wildness of the fields gave rise to superstitions, and a haunted house, from which groans proceeded and mysterious lights were seen at night, at one time scared the neighbourhood, and enticed the adventurous. The only historical incident, connected with the fields, is the duel fought there in 1829 between the Duke of Wellington and the Marquess of Winchelsea ; the latter having personally attacked the Duke during the debates on the Catholic Emancipation Bill.   The Duke aimed his shot through his adversary's hat, who then fired in the air, and the affair of honour was thus settled. Battersea Fields were approached, in those days, by the old wooden Battersea Bridge which had superseded the ferry ; the only means of communication till 1772. The present bridges at either corner of the Park have both been built since the Park was formed.

Like Victoria Park, Battersea was administered with the other Royal Parks, in the first instance.   The Act of Parliament giving powers to the "Commissioners of Her Majesty's Woods" to form the Park was passed in 1846, but so much had to be done to the land, that the actual planting did not begin until 1857.   The ground had all to be drained, and raised, and a proper embankment made to keep out the river. Just at this time the Victoria Docks were being excavated, and the earth dug out of them was conveyed to Battersea.   Places were left, to form the shallow artificial lake, mounds raised, to make the ground round the water undulating, and the rest of the surface of the Park levelled.   Altogether about a million cubic yards of earth were deposited in Battersea Park.   The extent is 198 acres, and from the nature of the ground, except the artificial elevations near the lake, it is quite flat.   The design was originally made by Sir James Pennethorne, architect of the Office of Works, and the execution of it completed by Mr. Farrow. The chief features, are the artificial water (for the most part supplied by the Thames), and the avenue of elms which traverses the Park from east to west, and cross walks, with a band-stand and drinking-fountain at the converging points.   Round the Park runs a carriage drive, and, following a different line, a track for riders—with the usual spaces for games between.   The trees are growing up well, so already any bareness has disappeared.   The absolute flatness, which makes the open spaces uninteresting, is relieved by the avenue, which will some day be a fine one.

It is an object-lesson to show the advantage of avenues and shady walks, too often ignored by modern park designers, or only carried out in a feeble, half-hearted way.   The chief variation in Battersea Park was achieved by John Gibson, the Park Superintendent, who made the sub-tropical garden in 1864.   His experience, gained on a botanical mission to India, which he undertook for the Duke of Devonshire, well fitted him for the task.   This garden has always been kept up and added to, and specially improved in the Seventies, while the present Lord Redesdale was at the Office of Works.

A sub-tropical garden was quite a novelty when first started here, and caused much interest to horti-culturalists and landscape gardeners.   The "Sub-tropical Garden," by W. Robinson, and other writings on the subject, have since made the effects which can be pro-duced familiar to all gardeners; but in 1864 to group hardy plants of a tropical appearance, such as aralias, acanthus, eulalias, bamboos, or fan palms, was a new idea.   During the summer, cannas, tobacco, various palms, bananas, and so on, were added to the collection, and caused quite an excitement when they first appeared at Battersea. The garden is still kept up, and looks pretty and cool in summer, and on a cold winter's day is sheltered and pleasant. But much of the charm and originality of the early planting has been lost, in the present official idea of what sub-tropical gardens should contain, which carries a certain stereotyped stiffness with it.

In 1887 the Park, at the same time as Victoria and Kennington, was given up to the Metropolitan Board of Works, and since then the control has passed to its successor, the London County Council. The gardens are kept up, more or less, as before, with a few additions. An aviary with a restless raven, fat gold and silver pheasants, and contented pigeons, delights the small children, who are as plentiful in Battersea as in all the other London playgrounds. Like the other parks, Saturdays and Sundays are the great days. The games of cricket are played as close together as possible, until to the passer-by the elevens and even the balls seem hopelessly mixed. The ground not devoted to games is thickly strewn with prostrate forms, and certainly, in this, Battersea is by no means singular! In autumn, one of the green-houses, in which the more tender sub-tropical plants are housed is given up to chrysanthemums. This flower is the one of all others for London. It will thrive in the dingiest corners of the town, and display its colours long after the fogs and frosts have deprived the parks and gardens of all other colour. The shows in the East End testify to what can be achieved, even by the poorest, with this friendly plant. Every year at Shoreditch Town

Hall the local exhibition takes place, and there are many similar institutions, where monster blooms, grown on roofs or in small back gardens, would compete creditably at a national show. The popularity of the chrysanthemums in Battersea Park is so great, that on a fine Sunday there is a string of people waiting their turn of walking through, stretching for fifty yards at least from the green-house to the entrance to the frame-ground. Certainly the arrangement of the green-house is prettily done. The stages are removed, and a sanded path with a double twist meanders among groups of plants sloping up to the rafters, and a few long, lanky ones trained to arch under the roof. The show is much looked forward to, and the colours and arrangements compared with former years, praised or criticised, such is the eager interest of those who crowd to take their turn for a peep. It is delightful to watch the pleasure on all faces, as a whole family out for their Sunday walk, press in together. It is only one more instance of the joy the London Parks bring to millions of lives.

The world of fashion has only attacked Battersea Park spasmodically. When it was new, and the sub-tropical garden a rarity, people drove out from Mayfair or Belgravia to see it. Again Battersea became the fashion when the cycling craze began. In the summer of 1895 it suddenly became " the thing " to bicycle to breakfast in Battersea Park, and ladies who had never before visited this South London Park flocked there in the early mornings. It was away from the traffic that disturbed the beginner in Hyde or St. James's Park, and perhaps the daring originality of cycling seemed to demand that conventions should further be violated; and nothing so commonplace as

Hyde Park would satisfy the aspirations of the newly-emancipated lady cyclists. What would their ancestors, who had paced the Mall in powder and crinolines, have said to the short-skirted, energetic young or even elderly cyclist? No doubt some of that language which shocks modern ears, used by the heroines in "Sir Charles Grandison," would have been found equal to the occasion. The great cycling rage is over, and Battersea is again deserted by fair beings, who now prefer to fly further afield in motors, but the Park is just as crowded by those for whose benefit it was really made—the ever-growing population of London south of the river.

Vauxhall Park

Going east from Battersea the next Park is Vauxhall, a small oasis of green in a crowded district. Although only 8 acres in extent, it is a great boon to the neighbourhood, and hundreds of children play there every day. It has been open since 1891, the land, occupied by houses with gardens, having been acquired and the houses demolished, and the little Park is owned and kept up by Lambeth Borough Council.

It has nothing to do with the famous Vauxhall Gardens, to which the rank and fashion of the town flocked for nearly two hundred years; and the country visitor to Vauxhall Park could hardly speak of it in such glowing terms as Farmer Colin to his wife in 1741 of the famous Vauxhall Spring Gardens:—

> "O Mary! soft in feature,
> I've been at dear Vauxhall;
> No paradise is sweeter,
> Not that they Eden call.

L

"Methought, when first I entered,
   Such splendours round me shone,
Into a world I ventured
   Where rose another sun."

The site of these Gardens, which covered some twelve acres with groves, avenues, dining-halls, the famous Rotunda and caverns, cascades and pavilions, is now all built over. It lay about as far to the south-east of Vauxhall Bridge as the little Park is to the south-west. In name Vauxhall sounds quaint and un-English. In earlier times it was known as Foxhall, or more correctly Foukeshall, from Foukes de Breant, who married a sister of Archbishop Baldwin in the latter half of the twelfth century.

The land of the present Park was purchased in May 1889.[1] Then it was covered by houses standing in their own grounds. The largest of these was Carroun or Caroone House, which had been built by Sir Noel de Caron, who was Ambassador of the Netherlands for thirty-three years, during the reigns of Elizabeth and James I.—the others, a row of eight with gardens, were known as "The Lawn." In front of them was a long pond, said to have been fed by the Effra River. This stream, which rose in Norwood and flowed into the

---

[1] As Vauxhall is not included in Lieut.-Col. Sexby's exhaustive book, the following details are not very accessible. It was bought from Mr. Cobeldick for £43,500.

Made up by Lambeth Vestry	.£11,746	17	6
„ Charity Commissioners . . .	12,500	0	0
„ London County Council . . .	11,746	17	6
„ Donations and other sources .	7,506	5	0
	£43,500	0	0

The fencing and laying out was done by the Kyrle Society. The Park was opened by the present King and Queen, July 7, 1890.

Thames at Vauxhall, has, like most of the other streams
of London, become a sewer, and the pond is no more.
In one of these houses (51 South Lambeth Road) Mr.
Henry Fawcett resided, and when the houses were pulled
down to form the Park his was left, the intention being
to make it into some memorial of him. It was found to
be too much out of repair to retain, and had to be pulled
down. With the sum which the sale of materials from
the old house realised, it was proposed to erect a memorial
drinking-fountain. This idea bore fruit, as Sir Henry
Doulton sold one to the vestry for less than one-third of
its value, and moreover gave a further memorial to the
courageous blind Postmaster-General of a portrait statue
by Tinworth, with appropriate allegorical figures.
This fine group recording the connection of Henry
Fawcett with the place is the most conspicuous feature
of the Park. The trees are growing up, and an abundance of seats and dry walks made it an enjoyable if not
beautiful garden. The swings and gymnasiums are
numerous and large, but what gives most pleasure is
the sand-garden for little children. For hours and hours
these small mites are happily occupied digging and making
clean mud pies, while their elders sit by and work. It is
touching to see the miniature castles and carefully patted
puddings at the close of a busy baby's day. In the
summer, when the sand is too dry to bind, some of the
infants bring small bottles, which they manage to get
filled at the drinking-fountain, and water their little hand-
fuls of sand. These children's sand-gardens, common in
parks in the United States, are a delightful invention for
the safe amusement of these small folk, and the delight
caused by this one, which was only made in 1905, shows
how greatly they are appreciated. Many of the parks

and some of the commons now have their "sea-side" or "sand-pit," and probably not only do they give immense pleasure, but they act as a safety-valve for small mischievous urchins, who otherwise could not resist trespassing on flower-beds.

The grass in this, as in all the parks, has to be enclosed at times, to let it recover, the tramp of many feet. The wattled hurdles which are often used in the London Parks for this purpose, have quite a rustic appearance. They are like those which appear in all the agricultural scenes depicted in fifteenth century MSS. It is much to be hoped that no modern invention in metal will be found to take their place.

Kennington Park

Not very far from Vauxhall, beyond the famous Oval, lies the larger and more pretentious Kennington Park of 19½ acres. This has a long history as Kennington Common. It formed part of the Duchy of Cornwall estates, having been settled by James I. on Prince Henry, and has since belonged to each succeeding Prince of Wales. In still earlier times there was a Royal Palace at Kennington, which fell into decay after Henry VIII.'s reign. Here as on all similar commons, the people had a right of grazing cattle for six months of the year. But the moment it was open to them in the spring such a number of beasts were turned on to the ground, that in a very short time "the herbage" was "devoured, and it remained entirely bare for the rest of the season."

The Common was a great place for games of all sorts, particularly cricket. When in 1852 it was turned into a Park, and play could not go on to the same extent, by suggestion of the Prince Consort, a piece of land, then market-gardens, was let by the Duchy to the Surrey Cricket Club, which was formed for the purpose of maintaining it. This is the ground that has since gained such notoriety as the Oval, the scene of many a match historical in the annals of cricket. The Common, too, was famous for the masses that collected there to hear Whitfield preach. His congregations numbered from 10,000 to 40,000 persons, and his voice would carry to the "extremest part of the audience." He notes in his diary, Sunday, May 6, 1731—"At six in the evening went and preached at Kennington; but such a sight I never saw before. Some supposed there were above 30,000 or 40,000 people, and near fourscore coaches, besides great number of horses; and there was such an awful silence amongst them, and the Word of God came with such power, that all seemed pleasingly surprised. I continued my discourse for an hour and a half." The last time he preached there was a farewell sermon before he went to America in August 1739.

Two other incidents are connected with Kennington Common, neither so pleasant—the scenes of the execution for high treason, with all the attendant horrors, of the "Manchester rebels" after the '45; and the great Chartist revolutionary meeting under Feargus O'Connor in 1848. The precautions taken by the Duke of Wellington saved the situation, and the 200,000 people who it had been proposed should march to Westminster melted away, and the whole thing was a fiasco.

It was soon after this episode that the Common was converted into a Park. The ground, including all the

Common and the site of the Pound, was handed over by the Duchy of Cornwall (by Act of Parliament), to be laid out as " Pleasure grounds for the recreation of the public; but if it cease to be so maintained " to "revert to the Duchy."

The transformation has been very successful, and the design was suitable and well conceived. The large greens are divided by wide paths shaded by trees, and each section can be closed in turn to preserve the grass. There is a sunk formal garden, bedded out with bright flowers, which show up well on the green turf; and at one end there are shrubberies with twisting walks in the style that is truly characteristic of the English Park, and seems to appeal to so many people. The whole space is not large, but the most is made of it, and both the formal and the "natural" sections have their attractions. At the "natural" end, near the church—which, by the way, was built as a thank-offering after Waterloo—is a handsome granite drinking-fountain, designed by Driver, and presented by Mr. Felix Slade; and in the centre of the Park is a fountain, given by Sir Henry Doulton, with a group of figures by Tinworth, emblematic of "The Pilgrimage of Life." The Lodge was the model lodging-house erected by the Prince Consort in the Great Exhibition of 1851.

MYATT'S FIELDS

Myatt's Fields or Camberwell Park is but a short distance to the south-west of Kennington. This Park of 14½ acres was one of those princely gifts which have been showered on the inhabitants of London. It was presented by Mr. William Minet, in whose family the land has been since 1770. His ancestors were

Victoria Manners. 1906

FOUNTAIN BY TINWORTH, KENNINGTON PARK

Huguenots who had come to England at the time of the Revocation of the Edict of Nantes.

It was handed over to the newly-formed County Council in 1889, having been previously laid out. The way in which this was done with an avenue, which will some day be one of the great beauties of the neighbour-hood, and which is in the meantime a pleasant shady walk, has already been commented on. For its size, Myatt's Fields is one of the most tasteful of the new parks. Its quaint name is a survival of the time when the ground was a market-garden leased by a certain Myatt from 1818–69. The excellent qualities of the strawberries and rhubarb raised there, gave the Fields such a good reputation in the district, and the name became so familiar, that it was retained for the Park.

Camberwell Green is a distinct place, not far distant, and is noticed among the village greens of London.

RUSKIN PARK

Ruskin Park, the newest of all the parks, is not very far from Camberwell, and has been formed of a cluster of houses, with grounds of their own, on Denmark Hill, known as the Sanders' Estate. The name, which has an "Art Nouveau" sound about it, and raises an expectation of something beautiful, was given to it because John Ruskin for many years lived in the neighbourhood. From 1823, when he was four, to 1843, his home was 28 Herne Hill, and there he wrote "Modern Painters." From then until 1871 he lived even nearer the present Park, at 163 Denmark Hill. Describing the house, Ruskin wrote of it: "It stood in command of seven acres of healthy ground . . . half of it meadow sloping to the sunrise, the rest prudently and pleasantly divided into an upper and lower kitchen-garden; a fruitful bit of orchard, and chance inlets and outlets of wood walk, opening to the sunny path by the field, which was gladdened on its other side in springtime by flushes of almond and double peach blossom." Such might have been the description of the houses and grounds now turned into a park. Some of the lines of the villa gardens have been retained, and some wise and necessary additions and changes have been made to bring the whole together; but even the inspiration of Ruskin has not kept out the inevitable edges and backbones of uninteresting evergreens. Some of the green-houses have been kept, but six dwellings have been demolished, and one of the two retained will be used as a refreshment room. The outside wall of the garden front of one, covered with wistaria, has been left, facing its own little terrace and lawn and cedars, and soon after the opening, in February 1907, many people found it was possible to get sun and shelter and enjoy the prospect from the seats in front of the ruined drawing-room windows. The dividing wall of two houses has been cleverly turned into what will be a charming pergola, and below, the ground has been levelled to form a bowling-green. The terraces and steps from one level to another are a pleasing feature in the design. The ground is not yet finished, and it is greatly to be hoped that the usual clumps of evergreens will not be multiplied, but Ruskin's description borne in mind, and let there be almonds and double peaches to gladden the spring, and not drooping, smutty evergreens, or "ever blacks," as they might be more fittingly called, to jar on the picture of fresh young growth. The pond, a stiff oval, has had to have the necessary iron railings, and the trees near it have been substantially barricaded with rustic seats—a most important addition. The avenue of chestnuts which crosses the open part of the ground has been left; and there are other good young trees growing up, and a fine old ilex and mulberry. There is already a question of adding a further 12 acres to this Park, which is 24 acres at present, but the scheme is still under consideration.

BROCKWELL PARK

Those who want a change, from the roar and bustle of streets, can attain their object very quickly by the expenditure of a few pence and fifteen minutes in the train. Getting out at Herne Hill Station, in a few seconds the gates of Brockwell Park are reached. The old trees and undulating ground are all that could be desired, but the chief attraction, and the object that well repays a visit, is the old walled garden. It is a high brick enclosure, with fine old trees peeping above, and festoons of climbing plants brightening the dull red walls. The narrow paths, running in straight lines round and across, are here and there, spanned by rustic arches covered with roses, or clematis, or gourds, from which hang glowing orange fruit in autumn. In the centre of the garden a small fountain plays on to moss-grown stones, and on a hot summer's day the seats, shaded by the luxuriant Traveller's Joy, make a cool resting-place, though not so sequestered as the arbours in the angles of the wall, darkened by other climbers. The rest of the garden is a delightful tangle of herbaceous plants. All the old favourites are there, and a small notice near the entrance announces to those in search of knowledge that the garden contains all herbs and garden

plants mentioned in Shakespeare's works. A little knowledge is a dangerous thing, and the unwary might not
realise that the flowers of Shakespeare's time, although
undoubtedly there, only form a small portion of the
whole display. The board is literally true, but visitors
are apt to go away with the idea that brilliant dahlias,
and gaudy calceolarias, or even the most modern introduction, *Kochia tricophila*, were friends of Shakespeare's!
A large number of the plants, however, are truly of the
Elizabethan age, that golden time of progress in gardening as well as of other arts, when spirited courtiers and
hardened old sailors alike scoured the seas and brought
strange plants from new lands. Many of these now
familiar treasures from east and west flourish in this little
enclosure, and recall the romantic days of the sixteenth
century: the Marvel of Peru—the very name tells the
delight that heralded its arrival from the West—the
quaint Egg-plant (*Solanum ovigerum*) brought from
Africa, and the bright-seeded Capsicums from India.
Even the bush, with its wealth of white or purple flowers,
the *Hibiscus Syriacus*, was known in those days, though
not by that name. Gerard, in describing it, says it was
a stranger to England; "notwithstanding, I have sowen
some seedes of them in my garden, expecting successe."
That delightful confidence, which is the great characteristic of all these old gardeners, was not abused, apparently, in this case, for two years later, in the catalogue
of plants in his garden, 1599, this great tree mallow was
flourishing. Many of the gourds, which are grown to
great advantage in this little garden, were also known at
an early date. Gerard says of them, "they joy in a
fruitful soil, and are common in England." Were it not
for the conspicuous little notice-board, no fault could be

found with the selection of plants which, from early spring till late autumn, brighten this romantic little garden. The *Solanum jasminoides* is none the less graceful because it has only found a home in sheltered corners in England, for the last seventy years. *Cobæa scandens*, which festoons very charmingly some of the arches, is certainly an old friend, having been over a hundred years in this country; but it is a new-comer when compared with the Passion Flower growing in profusion near it, and even that did not appear until after Shakespeare's death. It was unknown to Gerard, but his editor, Thomas Johnson, illustrates it in the appendix to the edition of 1633. It had then arrived from America, "whence it hath been brought into our English gardens, where it growes very well, but floures only in some few places, and in hot and seasonable yeares: it is in good plenty growing with Mistresse Tuggy at Westminster, where I have some years seene it beare a great many floures." Mistress Tuggy and her friend would have rejoiced at the sight of the house in the centre of Brockwell Park on a warm October day, thickly covered with the golden fruit as well as star-like flowers of their precious "Maracoc or Passion-floure."

This delightful walled garden was the old kitchen-garden. Luckily, the fashion for the gardens of a past generation was growing at the time the Park was purchased, and the London County Council must be congratulated on the good taste displayed in dealing with it. The history of the acquisition of the ground is soon told. The desire for a park in this neighbourhood led those interested to try and arrange to buy Raleigh House in the Brixton Road, with some 10 acres of land, for about

OLD ENGLISH GARDEN, BROCKWELL PARK

£40,000. Having got an Act of Parliament to allow this, Brockwell Park came into the market with a ready-made park of 78 acres. The Act of 1888 was repealed, and eventually a sum of nearly £120,000 was spent on the purchase of Brockwell, which was opened to the public in 1892. Near the entrance gates, close to Herne Hill Railway Station, a drinking-fountain, with a graceful figure of "Perseverance" and portrait bust, has been erected to Mr. Thomas Lynn Bristowe, M.P. for Norwood, who was chiefly instrumental in obtaining the Park, and whose death occurred with tragic suddenness at the opening ceremony. It is quite a steep hill up to the house, which is of no great antiquity or beauty, having been built at the beginning of last century, when the older manor-house was pulled down, by Mr. Blades, the ancestor of the last owner. The view on all sides is extensive, and the timber is fine. There are good old oaks, as well as elms and limes; and it is satisfactory to see that, in the recent planting, limes have been given a place, and not only the overdone plane. As a contrast to the delightful formal garden, some pretty wild grouping has been carried out beside the artificial water. This series of ponds are an addition to the Park as originally purchased. It now measures 84 acres, and the extra piece contained water, which has been enlarged into a big bathing-pool and a so-called "Japanese garden." These ponds are well arranged; and although there are various kinds of ducks and geese and black swans, and concrete edges and wire netting are inevitable, they are not so aggressive as in many parks. In places tall plants have been put in behind the railings and allowed to hang over, to break the undue stiffness. In the late autumn purple Michaelmas daisies nearly touched the water, and the red berries of the Pyracantha overhung the ducks without apparent disagreement.

The opening of Brockwell as a public Park has had the effect of banishing most of the rooks. There was a large rookery, but year by year the nests decrease. In 1896 there were thirty-five nests, the next year twenty, while in 1898 there were only eight or ten. Thus every season they are getting fewer, but still, in the spring of 1907, one pair of rooks were bold enough to build.

DULWICH PARK

Dulwich Park is not very far from Brockwell, but its surroundings are more open. A few of the roads near it have some feeling of the country left. The houses that are springing up are of a cheerful villa type, and have nothing of the monotony and dulness of most of the suburbs. Fine old trees grow along many of the roads. The chestnuts, for instance, in Half Moon Lane between Herne Hill and Dulwich are charming, and also on the further side of the Park, where the celebrated inn, the "Green Man," was situated, there is a rural aspect and a delightful walk between trees. It was within the grounds of the "Green Man" that the Wells of chalybeate water were situated. The Wells had been discovered in the reign of Charles II., and the water sold in London, but the "Green Man" did not become a popular resort until after 1739. A story connected with this popular spa is recorded in the "Percy Anecdotes" in 1823. A well-known literary man was invited to dinner there, and wished to be directed. However, he inquired vainly for the "Dull Man at Greenwich," instead of the "Green Man at Dulwich." One of the entrances to the Park is close

to the site of the once famous Wells.   The Park itself,
which covers 72 acres, was the munificent gift of
Dulwich College.   The gift was confirmed by an Act
of Parliament in 1885, and the Park opened to the
public in 1890.   The College was founded by Edward
Alleyn in 1614, who called it "The College of God's
Gift."   Originally, there were besides the Master,
Warden, and four Fellows, six poor brethren and six
sisters, and thirty out-members.   The value of the
property has so enormously increased that the number
of scholars has been very greatly added to, and now
hundreds of boys, some quite free, and some for a very
low fee, obtain a sound commercial education.   The
founder was a friend of Shakespeare, and one of the
best actors of his plays in the poet's lifetime.   His
early biographers go out of their way to refute the
alleged reason of his founding "God's Gift College,"
namely, that when on one occasion he was personating
the devil, the original appeared, and so frightened him
that he gave up the stage to devote himself to good
works.   Were this story true, the vision was certainly
well timed, and has produced unexpected and far-reach-
ing results.   The educational work, the picture gallery,
and the well laid out estate of Dulwich Manor, including
the large public Park, are all the direct result!

There are a few fine old trees in the Park, particularly a row of gnarled oaks near the lake.   This is a
small sheet of water on the side nearest the College.   The
carriage road, which encircles the Park, crosses by a
stone bridge the trickling stream, formed by the overflow from the lake.   On the south-east side of the Park
there are but few trees, but large masses of rhododendrons and azaleas have been planted, which make a

brilliant show in the summer. The most distinctive feature is the rock gardening. There is a very large collection of Alpine and rock plants, which are growing extremely well and covering the stones with delicious soft green cushions, which turn to pink, yellow, white, and purple, as the season advances. Even in the cold, early spring, snowdrops, and the pretty little Chionodoxa, the "Glory of the Snow," begin to peep out amongst the rocks, and these are the harbingers of a succession of bloom, through the spring and summer months. On either side of one of the entrances, a long and pleasing line of this rock-work extends, but the plants for the most part are grown on mounds like rocky islands rising up from a sea of gravel. There are several of these isolated patches in the middle of the carriage drive. It is certainly fortunate, for those who only drive round the Park, thus to have a full view of the charming rock plants; but to compare such a display to the rock garden at Kew is misleading. There may be nearly as many plants at Dulwich as at Kew, but the arrangement of that charming little retired valley at Kew is so infinitely superior that the comparison is unjustified. The small stream which leaves the lake, and other places in the Park, offer, just as good a foundation for a really effective rock garden as the one at Kew. Such an arrangement would give a much better idea of the plants, in their own homes, than the islands in the roadway, that must suffer from dust, besides looking stiff and unnatural. It is, however, delightful to see how well these plants are thriving. This is hardly astonishing, as it is not in a crowded, smoky district, but in one of the most favoured of suburbs. Dulwich Park adds greatly to the advantages of the neighbourhood:

it has not hitherto been crowded, and is by no means a playground of the poorest classes, but now the advent of electric trams and rapid communication may somewhat lessen its exclusiveness.

HORNIMAN GARDENS

There are gardens of a very different character round the Horniman Museum, not far distant. This collection, as well as the $9\frac{1}{4}$ acres of ground adjoining it on Forest Hill, were the gift of the late Mr. J. F. Horniman, M.P., and the garden, kept up by the London County Council, was opened in June 1901. The situation is extremely attractive. A steep walk up an avenue from London Road, Forest Hill, near Lordship Lane Station, leads to a villa standing in its own grounds, which is utilised for refreshment rooms and caretaker's house, &c. The lawns descend steeply on three sides, and on the western slope there is a wide terrace, with a row of gnarled pollard oaks. From this walk there is a wide and beautiful view, over the hills and parks, chimney-pots and steeples of South London, with the lawns and pond of Horniman Gardens in front. On this terrace a shelter and band-stand have been put up, and no more favoured spot for enjoying the open-air town life, so common on the Continent, but until lately so rare in England, can well be imagined. The country round is still fairly open, between Forest Hill and Brixton. Near the foot of Horniman Gardens lies Dulwich Park, with the shady path known as "Cox's Walk," from the proprietor of the "Green Man," and the roads lined with trees connect Dulwich with Brockwell Park, Herne Hill, so that this corner of London is well supplied with trees.

M

DEPTFORD PARK

Deptford Park is a complete contrast to the semi-rural Dulwich. It is in one of the most densely-populated and poor districts, where it is greatly needed, and has been open since 1897. The site was market-gardens, and was sold by the owner, Mr. Evelyn, below its value, to benefit the neighbourhood. It is merely a square, flat, open space of 17 acres, with only a few young trees planted round the outskirts. Near the principal entrance in Lower Road, the approach is by a short walk between two walls. Along either side of the pathway, and for some little distance to the right and left, after the open space is reached, a nice border of herbaceous plants has been made along the wall, and a few beds placed in the grass on either side, and ornamental trees planted. Thus the entrance to this wide playground is made cheerful and attractive, and a pleasant contrast to the grimy streets outside.

TELEGRAPH HILL

Between these two extremes lies a small Park known as Telegraph Hill. It is only 9½ acres, and is cut in two by a road, but it is very varied in surface. The origin of its name is from its having been a station for a kind of telegraphy that was invented before the electric telegraph had been discovered. Two brothers Chappé invented the system, and were so successful in telegraphing the news of a victory in 1793, that their plan was adopted in France, and soon throughout Europe. In Russia a large sum was expended in establishing a line of communication between the German frontier and St. Petersburg; but so slow was the building that the stations were hardly at work before they were superseded by electricity. The signals were made by opening and shutting six shutters, arranged on two frames on the roofs of a small house, and by various combinations sixty-three signals could be formed. The Admiralty established the English line, of this form of telegraphy between Dover and London in 1795, and the first public news of the battle of Waterloo actually reached London by means of the one on "Telegraph Hill." The place was well chosen, for even now, all surrounded by houses, the hill is so steep and conical, that a very extensive view is still obtained. The site of the semaphore station is now a level green for lawn tennis. On the other side of the roadway, the descent is steep into the valley, and there are two small ponds at the bottom. The cliffs are covered with turf, interspersed by the usual meaningless clumps of bushes, and a few nice trees.

Southwark Park

Southwark Park lies far away from Southwark, beyond Bermondsey, in Rotherhithe. It was in the parliamentary borough of Southwark, hence the misleading name. The Park is a gloomy enough place when compared with the more distant or West End Parks, but a perfect paradise in this crowded district. Between its creation in 1864 and its completion in 1869, a great reformation was worked in the district. Close to the docks, and intersected by streams and canals, with the poorest kind of rickety houses so vividly described by Dickens in "Oliver Twist," the surroundings were among the most dismal imaginable. The actual site of the Park was partly market-gardens, which had for long been established in this locality owing to the fertility of the alluvial soil.

Vines were grown here for wine with success in the first half of the eighteenth century, when there was a revival in grape-growing, and vineyards were planted at Hoxton and elsewhere. Over 100 gallons of wine were made in a year in Rotherhithe. Some of the earth excavated from the Thames Tunnel was put on the ground covered by the Park before the laying out commenced. When the land, 65 acres, was bought, only 45 were to be kept for the Park, and the rest were reserved for building. But when the day of building arrived there was such an outcry that the whole plan was remodelled, the drives which encircled it done away with, and tar-paved paths substituted, only one driving road crossing it being left, and the ponds added. It is more the want of design, than any special style, that is conspicuous, and a good deal more could have been done to make the Park less gloomy. An avenue is growing up, but it will never have the charming effect of the one across Battersea, as the line is neither straight nor a definite curve. The wild fowl on the pond are such an attraction, that perhaps it may be that the wire netting and asphalt edges they apparently require are not drawbacks, but they are not beautiful. The gateway into the Park, near Deptford Station, has rather the grim look of a prison, and yet, with the forest of masts behind, all it requires is a climbing plant or two to make a picture. On the opposite end of the Park runs Jamaica Road, which perpetuates the name of a well-known Tea Garden, Jamaica House. Pepys records a visit there, on a Sunday in April 1667. "Took out my wife, and the two Mercers, and two of our maids, Barker and Jane, and over the water to Jamaica House, where I never was before, and there the girls did run for wagers over the bowling-green;

and there, with much pleasure, spent little and so home."
Pepys' home in Seething Lane near the Tower would be
an easy distance from the Tea Gardens of Redriff, as
Rotherhithe was called then, and in the days when Swift
made Gulliver live there. There were other well-known
Tea Gardens near, the "Cherry Garden," "Half-way
House," and at a much later date "St. Helena's Gardens,"
which were only closed in 1881. The disappearance of
all the Tea Gardens and open spaces made the necessity
of a Park very obvious, and it was to meet this want that
Southwark Park was made.

MARYON PARK

There is one more small Park to complete the line of
South London Parks, for which the public is indebted to
Sir Spencer Maryon-Wilson, the lord of the manor of
Charlton, in which parish it is situated. It lies between
Greenwich and Woolwich, and the South-Eastern Railway
skirts the northern side. The ground was chiefly large
gravel pits, and has a hill in the middle partly caused by
the excavations. This hill has some pretty brushwood
still growing on its slope, showing it was once joined to
Hanging Wood, a well-known hiding-place of highwaymen. It was conveniently thick, and there are many tales
of pursuit from Blackheath which ended by losing the
thieves in Hanging Wood. The hill in the Park is
locally known as Cox's Mount, having been rented by
an inhabitant of that name in 1838, who built a summer-
house there and planted poplars. The area of the Park
is about 12 acres, and except for one or two trees on the
Mount and patches of brushwood, it is open grass. The
boys on the *Warspite* training ship anchored near are
allowed to play cricket there, provision for this having

been made by the generous donor of the Park in the deed of gift to the London County Council in 1891.

Quite outside these crowded districts, yet within the County of London, lie three more Parks maintained by the County Council. The one nearest the heart of London is Manor Park, or Manor House Gardens, between the High Road, Lee, and Hither Green Station, opened in 1902. There are 8¾ acres here attached to the Lee Manor House, a substantial building in the Adams style, now used as the Public Library. The Gardens slope gently away from the house to a large pond—or lake as the Council would prefer to call it—and beyond to a rapid little stream, the Quaggy, a tributary of the Ravensbourne. Beyond the Quaggy's steep banks, well protected by spiked railings, is a flat green devoted to games. The chief beauty of this little Park is four magnificent old elms and a few other good trees—beech, chestnut, *Robinia speudo acacia*, &c. In the spring of 1907 the pond was in process of cleaning, so no rooks had ventured to build within the Park, but just at the gates a large elm in a small garden had been favoured by these capricious birds, and their hoarse voices were making a deliciously countrified sound.

The other London County Council Parks are in what is still nearly open country, although rows of villas are being rather rapidly reared in the district. Eltham is one of these. It is at present not enclosed with massive iron railings, but the wide, flat stretch of smooth turf, studded with patriarchal trees, is left untouched, except that a few spaces have been levelled for games. This Park of 41 acres was bought in 1902, the Borough of Woolwich paying half the cost of purchase—£9600—with the Council.

Still further into the country is Avery Hill, with

the large house and grounds, extending over 84 acres, built and laid out by Colonel J. T. North. The London County Council were offered this estate in 1902, if purchased within a certain limit of time, for £25,000. Usually the Council, in making a purchase, have ascertained beforehand what contributions the local Boroughs were prepared to subscribe towards the total cost, but, on this occasion, the Boroughs were invited to share the expense after the purchase had been made, with the result that all those concerned—Camberwell, Lewisham, Greenwich, Deptford, and Woolwich—refused; so the whole of the purchase and upkeep devolved on the London County Council. The large mansion is now used as a teachers' training college for girls, but the greater part of the grounds, and the immense winter gardens are open to the public. It is still so far from the centres of population that the public who make use of these spacious gardens is very limited. The nearest railway station, New Eltham, is three-quarters of a mile distant from the Park, and half-an-hour or more by train from Charing Cross. Although it is now so far into the country, and some people would deprecate the purchase, it is only fair to remember that most of the crowded districts were also country not long ago, and that when land is dear and houses being built is not a favourable moment to purchase. As a rule it is want of foresight that is the complaint, and not excess of zeal, as in this case. The garden is made use of to furnish supplies of plants to some of the smaller parks, and a portion is being reserved for growing specimens for demonstration in the Council Schools. On the west side of the house there are three terraced gardens, prettily planted with roses and fruit-trees. In front of the house a sloping lawn, with a few large beds, touches the park-like meadows studded with trees. Sheep feeding with their tinkling bells gives a rural appearance. To the large, modern, very red brick house is attached a huge winter garden. This is on a very large scale, with lofty palms, date, dom, and cocoa-nut growing with tropical luxuriance in the central house, with a large camellia house on one side and a fernery with rock-work, pools, and gold-fish on the other. All this requires a good deal of keeping up—nearly £3000 a year—and although it has been open now some five years, it has been enjoyed by few. It is greatly to be hoped that it has a much-appreciated future before it.

Such is a slight sketch of some of London's Parks. No doubt there is much that could be changed for the better, both in design and planting: less sameness and meaningless formality without true lines of beauty in design would be an improvement. In planting, there might be more variety of British trees—alder, oak, ash, and hawthorn; and a wider range of foreign ones—limes, American or Turkey oaks, and many others; more climbing plants, such as Virginian creepers, more simple herbaceous borders and fewer clumps of unattractive bushes, and more lilacs, laburnums, thorns, almonds, cherries, and medlars in groups on the grass. If greater originality was displayed and a thorough knowledge of horticulture were shown, especially by the authorities that supervise the largest number of these parks, many improvements in existing ones could be easily achieved, and in forming new parks the same idea need not be so rigidly followed. But, in spite of small defects, the Parks as a whole are extremely beautiful, and Londoners may well be proud of them.

CHAPTER VIII

COMMONS AND OPEN SPACES

'Tis very bad in man or woman
To steal a goose from off the common,
But who shall plead that man's excuse
Who steals the common from the goose ?
—An Old Ditty.

IT was only fifty years ago, when the want of fresh air and room for recreation was being realised, that people began to wake up to the truth that there were already great open spaces in London which ought to be cared for and preserved. It was brought home by the fact that over £1000 an acre was being paid to purchase market-gardens or fields so as to transform them into parks, while at the same time land which already belonged to the people was being recklessly sold away and built over. All through the history of most of the common lands encroachments of a more or less serious nature are recorded from time to time. The exercise of common rights also was often so unrestrained as to inflict permanent injury on the commons. The digging for gravel was frequently carried to excess, whins and brushwood were cut, and grass over-grazed until nothing remained. At last, in 1865, a Commons Preservation Society was formed with the view of arousing public attention to the subject. As is often the case, some people ran to the opposite extreme, and wished to transform the commons into parks without giving compensation to the freeholders and copyhold tenants, who thereby would lose considerable benefits. In some cases after the Metropolitan Commons Act of 1866 was passed, the Lord of the Manor, on behalf of all the freeholders, disputed the right of the Metropolitan Board of Works to take the land without compensation to the owners. The lord of the manor was considered unreasonable by some of the agitators for the transference of the common lands to public bodies, but he was fighting the battle of all the small owners. The freeholders in some cases were as many as fifty for some 40 acres. Many of the commons were Lammas Lands. The freeholders, of which there were a large number, had the use of the land from the 6th of April until the 12th of August, and the copyhold tenants of the manor had the right of grazing during the remainder of the year. The number of cattle each could graze was determined by the amount of rent they paid, and the grazing was regulated by the " marsh drivers," men elected annually by the courts of the Manor for the purpose. A curious incident in connection with these rights happened on Hackney Downs in 1837. The season was late, and the steward of the Manor put up a notice to the effect that as the freeholders' crops were not gathered the grazing on the Downs could not begin until the 25th, instead of the usual 12th of August. The marshes and other common lands in the parish were open, so there was actually plenty of pasture available for those entitled to it. There was a fine crop of wheat on some plots on the Downs, and on the morning of Monday the 14th August, " a few persons made their appearance and began to help themselves to the corn." Summoned before the magistrates, the bench decided that after the usual opening day the corn "was common property, and could be claimed by no one parishioner more than another." On the strength of this decision the whole parish turned out, and a terrible scene of looting the crop took place, while the poor owners vainly tried to save what they could. The freeholder with the most wheat, a Mr. Adamson, lost over £100 worth, although he worked all night to save what he could. A case followed, as Mr. Adamson prosecuted Thomas Wright, one of the many looters who thought they had a right to it, for stealing his wheat. This time the magistrates fined the man twenty shillings, and half-a-crown, the value of the wheat he had actually taken, as he had no right to take away the crop, although he had a right to put cattle on the Downs. Further trials for riot before the Court of Queen's Bench resulted in the prisoners being discharged after they had pleaded guilty. It appeared both the looters and Mr. Adamson were in the wrong. They had no right to remove the corn, neither had he, after the 12th August, and those who had grazing rights could have turned on their cattle to eat the standing corn. This incident just shows how the right of freeholders and copyholders could not lightly be trifled with.

The report of the Select Committee on Open Spaces in 1865 pointed out in the same way, that although the right to these common lands had been enjoyed from time immemorial, the rights were vague as far as the public at large were concerned. They were probably limited to a certain defined area or body of persons, as the inhabitants of a parish, and it was doubtful if the custom would hold good at law for such a large place as London. Thousands of people from all parts of London trampling over a common was a very different thing to the free use of it by the parishioners. This report led to the passing of the Metropolitan Commons Act of 1866. Both before and after this Act there were several others for the maintenance and regulation of the commons and all the parks, gardens, and open spaces too numerous to mention.[1]

Under the present system most of the metropolitan commons and heaths are in the hands of the County Council, and in some cases considerable sums have been spent on them. Among the smaller ones is London Fields, Hackney, the nearest open space to the city. This was in a very untidy state when first taken in hand after 1866. The grass was worn away, and it was the scene of a kind of fair, and the resort of all the worst characters in the neighbourhood. It used to be known as Shoulder of Mutton Fields, and the name survives in a "Cat and Mutton" public-house on the site of a tavern which gave its name to the fields. It was in the eighteenth century a well-known haunt of robbers and foot-pads, and in spite of a watch-house and special guard robberies were frequent. The watch must have been rather slack, as about 1732 a Mr. Baxter was robbed about five in the morning "by two fellows, who started out on him from behind the Watch-House in the Shoulder of Mutton Fields." Hackney is rich in open spaces, as besides London Fields there is Hackney or Well Street Common, near Victoria Park, Mill Fields, Stoke Newington and Clapton Commons, Hackney Downs

---

[1] See "Chitty's Statutes," by J. M. Lely, under "Metropolis."

(over 40 acres) on the north, and Hackney Marshes (337 acres) on the east. These were Lammas Lands, and the marshes were used for grazing until within the last few years, when the rights were bought up and the land finally thrown open to the public in 1894. The river Lea skirts the marsh, and used not unfrequently to flood, doing considerable damage. The London County Council have made four cuts across the bends of the river, forming islands. The water now can more easily flow in a wet season, and the periodical inundations no longer occur. The planting of these islands has not been carried out at all satisfactorily. An utter want of appreciation of the habits of plants or the localities suited to them has been shown. A stiff row of the large saxifrage, *S. cordifolia*, charming in a rock garden or mixed border, has been put round the water's edge, and behind it, berberis, laurels, and a few flowering bushes suited to a villa garden shrubbery. The opportunity for a really pleasing effect has thus been missed, and money wasted. A few willows and alders, with groups of iris and common yellow flags, and free growing willow herb, and purple loosestrife, would soon, for much less expense, have made the islands worthy of a visit from an artist. Instead, an eyesore to every tasteful gardener and lover of nature has been produced. The beauty of the marsh has always been appreciated by the dwellers in Hackney and Clapton. The view over the fertile fields from the high land was one of the attractions since the time when Pepys wrote, "I every day grow more and more in love with" Hackney.

Hackney Downs now form a large open area for recreation, but they were fruitful fields sixty years ago. An engraving, from a drawing by W. Walker, dated

1814, represents a "Harvest Scene, Hackney Downs, with a View of the Old Tower, and Part of the Town of Hackney," and gives a delightful picture of harvesters reaping with sickles, and binding up sheaves of the tall, thick-growing corn. That some of the Downs were arable land was a grievance to those who had grazing rights, and there was a considerable agitation to get the freeholders to lay it all down in grass, after the incident of looting the corn in 1837, already referred to. The Downs continued rural within the memory of many still living. The Lord of the Manor remembers that an inhabitant stated that she had, whilst walking across the Downs, startled a wild hare from her form. This would be about the year 1845, and for ten or twelve years later there were partridges in the larger fields of turnip and mangold-wurzel which adjoined the Downs. The rural character has quite changed, and now the Downs are a large open space, with young trees growing up to supply shade along the roads which encircle the wide grassy area.

Highbury Fields, although much smaller than Hackney Downs, being only 27 instead of 41 acres, play as important a part in the north of London, as the Downs do in the north-east. They are not, however, Common Lands, but until recently were actually fields with sheep grazing in them. Tradition points to Highbury Fields as the site of the Roman encampment during the final struggle with Boadicea. In the Middle Ages they belonged to the Order of St. John of Jerusalem, and there the rebels of the Wat Tyler rising, headed by Jack Straw, camped after leaving Hampstead. There are a few old trees still standing in the Fields, which were formerly within the grounds of two detached residences, one of them the Manor House. An old "moated grange," or barn, belonging to the ancient Priory, gives its name to the public-house, Highbury Barn, the goal of motor omnibuses. The moat was only filled up fifty years ago, and the old buildings pulled down, after enjoying some notoriety as a Tea Garden for over a century. A part of the present Fields was called "the Reedmote," or "Six Acre Field," and is also shown on old maps as "Mother Field." When Islington Spa was a fashionable resort, and Sadler's Wells at the height of its prosperity, the houses facing the Fields were built. On the north-west the row is inscribed in large letters, "Highbury Terrace, 1789," and this, according to old guide-books, "commands a beautiful prospect." On the east lies another substantial row of eighteenth-century mansions, and the inhabitants are proud to point out to strangers No. 25 Highbury Place as the house in which Mr. Chamberlain lived, from the age of nine until he was eighteen, when he went to live in Birmingham. His present home, now so well known, was built in 1879, and was named in remembrance of Highbury Place. In the early years of the nineteenth century several well-known people were living in these houses. John Nichols, the biographer of Hogarth, who was for fifty years editor of the *Gentleman's Magazine*, died there in 1826. A few years later a historian of Islington describes Highbury Place as "thirty-nine houses built on a large scale, but varying in size, all having good gardens, and some of them allotments of meadow land in the front and rear The road is private, and is frequented only by the carriages passing to and from the several dwellings situated between the village and Highbury House." This description draws a very rural picture, of which nothing now remains but the name.   The Fields were turned into a public Park in 1885, and now consist of wide open spaces for games, with intersecting paths well planted with limes, elms, chestnuts, and planes, and an abundance of seats.   Near the point where Upper Street, Islington ends and Holloway Road joins it, a memorial to the soldiers and volunteers of Islington who fell in the Boer War has been erected,-and the figure of Victory stands conspicuously facing the approach from the city.

By far the most beautiful and the most frequented of all London Commons is Hampstead Heath.   The original Heath measured 240 acres, but, with the addition of Parliament Hill, there are now over 500 acres of wild open country for ever preserved for the benefit of Londoners.   'Appy 'Ampstead, the resort not only of 'Arrys and 'Arriets, but poets, artists, and people of every rank in life, is too well known to demand description.   The view from it seems more beautiful every time the occasional visitor ascends the hill, and gazes down on London and away over the lovely country of the Thames valley. The County Council, the present holders of this public trust, have mercifully refrained from turning it into a park—the original intention of those who first wished to preserve it.   The bracken still flourishes, the gorse still blooms, and there is yet a wild freshness about it that has not been "improved" away.

Hampstead has had periods of fashion as a residence. In the eighteenth century it is described as " a village in Middlesex, on the declivity of a fine hill, 4 miles from London.   On the summit of this hill is a heath, adorned with many gentlemen's houses. . . . The water of the [Hampstead] Wells is equal in efficacy to that of Tunbridge, and superior to that of Islington."   These Wells appear to have first attracted notice in the time of Charles II. In 1698, Susanna Noel and her son, third Earl of Gainsborough (then the owner of the soil), gave the Well, with six acres of ground, to the poor of Hampstead. For more than thirty years the Wells, with all the attendant attractions of the pump-room, with balls and music, drew the fashionable world up to Hampstead. It was said to be " much more frequented by good company than can well be expected, considering its vicinity to London ; but such care has been taken to discourage the meaner sort from making it a place of residence, that it is now become . . . one of the Politest Public Places in England." Here Fanny Burney made her heroine, Evelina, attend dances, and it plays a part in the fortunes of Richardson's Clarissa Harlowe ; and here all the wits and poets of the time mingled in the gay throng. Many have been the celebrated residents in Hampstead—Lord Chatham, Dr. Johnson, Crabbe, Steele, Gay, Keats, William Blake, Leigh Hunt, Romney and Constable, John Linnell, and David Wilkie among the number. The site of the pump-room is all built over, but some fine old elm trees in Well Walk, still have an air of romance and faded glory about them. The houses near the Heath—such as Shelford, afterwards Rosslyn House, with a celebrated avenue of Spanish chestnuts, The Grove, Belsize Park, the residence of Lord Wotton, and then of Philip, Earl of Chesterfield—have all been consumed by the inroads of bricks and mortar. It is more than likely that the Heath would have shared the same fate, had not the inhabitants taken active steps to arouse public attention to preserve this wild heath, unequalled near any great city. Already aggressive red villas were making their appearance in far too great

N

numbers. The western side was dotted over with them.
That the purchase of it for the public benefit has been
appreciated it is not difficult to prove, when over 100,000
visit it on a Bank Holiday. It was the commencement of
building operations near the Flagstaff by the lord of the
manor, Sir Thomas Maryon Wilson, in the heart of the
Heath, that brought things to a crisis in 1866. A case
began against the lord of the manor, but he died before
it was ended, and his brother, Sir John, being willing to
compromise, the sum of £47,000 was agreed on for the
sale of the Heath to the Metropolitan Board of Works.
The few houses dotted about on the Heath are those of
squatters, who have established their right by the length
of time they have been in possession. The small hamlet
or collection of houses in the " Vale of Health," those
near the " Spaniards " and round Jack Straw's Castle,
have existed from time immemorial, although few old
houses of interest remain, and large, unsightly buildings
have taken the place of the picturesque ones. In the
Vale of Health the houses are chiefly given up to catering
for holiday-makers. The " Spaniards," at the most
northerly point of the Heath, is a genuine old house,
and it still has a nice garden, although all the alleys
and fantastic ornaments which made it popular, in the
eighteenth century, have vanished. The name came from
the fact that the first owner was a Spaniard. The next
proprietor was a Mr. Staples, who "improved and beautifully ornamented it." The house was on the site of the
toll-gate and lodge to Caen Wood, and its position saved
that house from destruction, at the time of the Gordon
riots. The rioters had burnt and wantonly destroyed
Lord Mansfield's house in Bloomsbury Square. Mad-
dened with drink, and flushed with triumph at the success

of their outrages, they made a bonfire in the square of
the invaluable books collected by Lord Mansfield. Their
temper may be imagined as they marched by Hampstead
to commit the same violence at Caen Wood, Lord Mansfield's country house. The proprietor of the "Spaniards"
invited them in, and threw open his cellars to the mob.
Fresh barrels of drink were sent down from Caen Wood,
and meanwhile messengers were despatched for soldiers;
so that by the time all the liquor had been consumed, and
the drunken rioters began to proceed, they were confronted
by a troop of Horse Guards, who, in their addled condition, soon put them all to flight. The name of the
other inn on Hampstead Heath, which stands conspicuously on the highest point, 443 feet above the sea,
is Jack Straw's Castle, and has also some connection with
a riot. Jack Straw was one of the leaders in the Wat
Tyler rebellion, and after burning the Priory of St. John
of Jerusalem, he came up to Hampstead and Highgate,
though there is no direct evidence to connect him, in
1381, with any tavern on the spot on which the inn
stands. The addition of Castle to the name is from
the fact, that there was some sort of fortress or earth-
works on this commanding point. The inn on the
site was known as the Castle Inn, and not until 1822
is there any mention of it as Jack Straw's Castle. The
wood of the gallows on which a famous highwayman
was hung behind the house in 1673 was built into
the wall. Jack Straw's Castle is now quite modernised,
but the view from it, on all sides, is still as lovely as
ever. The Whitestone Pond in front is really a
reservoir, and to the south of that lies the Grove, with
fine trees and some old-fashioned houses. The most
picturesque walk is that known as the Judges' or

King's Bench Walk, from a tradition that justice was administered under the trees there, when the judges fled from London at the time of the Great Plague. This walk is on the south-west side of the Heath, the Well Walk on the south-east. To the east of the highest point with Jack Straw's Castle and the road which runs northwards towards the "Spaniards" is the Vale of Health, and below are a series of ponds. Hampstead has always furnished a water-supply for the city at its feet. When more water was required, in the sixteenth century, the Lord Mayor proposed to utilise the springs there, and convey the water to London by conduits. A pound of pepper at the Feast of St. Michael annually to the "Bishop of Westminster," was the tribute for the use of the water, as the land belonged to the Abbey of West-minster, having been granted to it by King Ethelred in 986. The managers of water-supply in 1692 were a company known as the Hampstead Water Company, which became absorbed in the New River Company. The lakes are very deep, and dangerous for boating, bathing, and skating, although used for all those purposes.

The hill which rises beyond the ponds and stretches away to the east, is part of the land adjoining the true Heath, which was bought in 1887, so as to double the area of open country, and prevent that side of the Heath being overlooked by houses. The character is quite a contrast, and lacks the wildness, but it is pretty, park-like scenery, and Hampstead Heath would have been greatly spoilt had this further wide space of pasture land not been saved. The first hill to the east of the Heath is crowned by a mound or tumulus, which was opened a few years ago ; the investigations leading scientists to believe that it was a British burial-place of the bronze age. This used to be very picturesque with a group of Scotch firs—now, alas ! all dead. The next hill is Parliament or Traitor's Hill, and there is no very definite solution of the name. It may have been a meeting-place of the British "Moot" or Parliament, or the origin may only be traced to Cromwell's time. As if to encourage the tradition being kept up, a stone suggests that meetings may take place within 50 yards of the spot by daylight. Below the hill are flat meadows by Gospel Oak, said to be so named from its being a parish boundary, and the Gospel was read under the tree to impress the parishioners, with the same object as the other and more familiar form of beating the bounds. These Gospel Oak fields are the typical London County Council greens for games, so gradually, after leaving the summit of the Heath, the descent is made, from the artistic and picturesque, to the practical and prosaic.

Hampstead was always famous for its wild flowers. The older botanists roamed there in search of rare plants, and the frequent references in their works, especially in Gerard's " Herbal," show how often they were successful. Osmundas, or royal ferns, sundew or drosera, and the bog bean grew in the damp places, and lilies of the valley were among the familiar flowers. As late as 1838 a work on London Flora enumerates 290 genera, and no less than 650 species, as found round about the Heath. The soil, the aspect, the situation, are all propitious. Even now it is so far above the densest smoke-fogs that much might be done to encourage the growth of wild flowers. It is true notice-boards forbid the plucking of them, and that is a great step in advance—but the sowing of a few species, which have become extinct, would add greatly to the charm of the place. It is also still the favourite haunt of wild birds, and the more the true wildness is encouraged, the more likely they are to frequent it. It is much to be hoped that the London County Council will refrain in their planting, from anything but native trees and bushes which look at home, and which would attract our native songsters. Within the last ten or twelve years a very great variety of birds have been recorded either as nesting there or as visitors. The following list (taken from "Birds in London" by W. H. Hudson, 1898) may interest bird lovers :—

Wryneck, cuckoo, blackcap, grasshopper, sedge, reed and garden warblers, both white-throats, wood and willow wrens, chiff-chaff, redstart, stonechat, pied wagtail, tree pipit, red-backed shrike, spotted fly-catcher, swallow, house martin, swift, goldfinch, wheat-ears in passage, fieldfare in winter, occasionally red-wings, also redpoles, siskin, and grey wagtail.

This list is certainly a revelation to those who only associate dusty sparrows and greedy wood-pigeons with the ornithology of London. No better testimony is wanted to prove that Hampstead is still the beautiful wild Heath that has given pleasure to so many generations.

The only other large space of common land, north of the river within the London area, is Wormwood Scrubs, of very different appearance and associations from Hampstead. The manorial and common rights were purchased by the War Office, and the ground made over to the Metropolitan Board of Works in 1879, with re-servations for the rifle range and military exercises. The space is altogether over 200 acres. The ground in ancient times was a wood, adjoining "Old Oak Common," just beyond the London boundary, which was covered with patriarchal oaks. The last was felled in 1830. The ground, being flat, is admirably suited for the War Office purposes; it has gone through a process of draining, and the only part not downtrodden by soldiers has been "improved" by the London County Council, so there is little wildness or attraction in the place. The presence of a prison, erected in 1874, still further diminishes its charm as an open space.

This completes the open large spaces on the north; the south of the river is even richer in commons. One of the most thoroughly rural spots within the London area is Bostall Wood. There is nothing to spoil the illusion, and for quite a considerable walk it would be easy to imagine that a journey on the magic horse of the "Arabian Nights" had been taken to some distant forest land, to forget that the roar of the town was barely out of one's ears, and that ten minutes' walk would take one, out of the enchanted land, back to suburban villas and electric trams.

Beyond the inevitable band-stand, which attracts thousands on a summer Sunday evening, there is nothing to jar, and spoil the illusion of real country. The woods, and Bostall Heath which adjoins them, can be reached from Plumstead or Abbey Wood Station, in twenty minutes' walk up the steep hill. Pine woods crest the summit, and below them stretches a delightful thicket, chiefly of oaks and sweet chestnut, with an undergrowth of holly and a pleasant tangle of bracken and bramble,

where the blackbirds, chaffinches, and robins call to each
other and flit across the path. Steep slopes, and valleys,
and hollows clothed with trees, give possibilities of real
rambles, in a truly sylvan scene. Under the pines, which
are tall enough to produce that soothing, soughing sound
even in the most gentle breeze, the carpet of pine needles
is cushioned here and there with patches of vivid green
moss where the moisture has penetrated. Beyond the
Wood lies the Heath, studded with birch trees, among
gorse and bracken. There are narrow gullies and glades,
like miniature " gates " or " gwyles " of the sea coast, and
at the foot of the Heath lie the marshes, often in the
soft light as blue as the sea, and the silver Thames, a
bright streak across the picture, chequered with the red
sails of the barges, and tall masts of the more stately ships.

The whole area of woods and common is only about
133 acres, but the varied surface, and the distant views
from it, make it appear of larger extent. It is little
known to most Londoners, although the Heath was purchased as far back as 1877, and the Wood bought by the
London County Council in 1891. The place, however, is
much frequented and duly appreciated by the neighbouring population. This peaceful country-side could be
reached within an hour, from any point in the City. It is
attractive at all times of the year, especially in spring,
when the green is pale and the young brackens, soft and
downy, are uncurling their fronds, and the dark firs stand
up in sharp contrast to the tender greens. Or, perhaps,
still more delightful is it in autumn, when

" Red o'er the forest gleams the setting sun,"

and the oaks have turned a rich russet, and the birches,
of brilliant yellow, shower their tiny leaves on the mossy

earth, like the golden showers which fell on Danaë in her prison.

The attractive wood-clad hills of Bostall are the most remote of all London's open spaces. They lie the furthest east on the fringe of the suburbs. From Bostall westward roofs and chimney-pots become continuous— Woolwich, Greenwich, Deptford, Bermondsey, Southwark getting more and more densely crowded. But westward also begins the chain of commons which circle the town round the southern border—with breaks, it is true, yet so nearly continuous that from the highest point of one, the view almost ranges on to the next.

Only a deep valley, with Wickham Lane on the track of a Roman road, divides Bostall Wood from Plumstead Common. This is open and breezy, standing high above what was in ancient times the marsh overflowed by the Thames. The greater part is, however, used by the military, and the trample of horse artillery makes it look like a desert. It is a curious effect to see this part of the Common in winter. It has probably been used for manœuvring all the week, and by Saturday afternoon there are pools of mud, and ruts, and furrows, and hoof-marks all over it. On this dreary waste hundreds of boys and young men, sorted according to age, play more or less serious football matches. The coats of the players, in four little heaps, do duty for goal-posts, and these are so thickly strewn over the surface, and the players so closely mingled, that the effect is like bands of savages fighting among their slain—the ancient barrow in the centre of the ground gives colour to the supposition.

A sudden deep valley, called " the Slade," cuts the Common in two. In the hollow there are ponds, and on the high ground beyond stood a windmill, the remains of which are embedded in the Windmill Tavern.

The next common west of Plumstead, is Woolwich, maintained by the War Office and given up to military exercises. The extent is 159 acres. It is so much absorbed by the requirements of the War Office that it cannot be classed among London's playgrounds.

Going westward, the next large space is Blackheath, whose history is wrapped up with that of Greenwich, the beautiful Greenwich Park having once been part of the Heath. It is high ground, for the most part bare of trees, and with roads intersecting it—one of them, the old Roman Watling Street. The wild, bare summit of the Heath was a dangerous place for travellers, and many was the highway robbery committed there in times past. It is of very large extent, some 267 acres, and has been effectually preserved for public use, for some thirty-five years, since early in the Seventies.

The Heath has played its part in history—gay scenes, such as when the Mayor and aldermen of London flocked, with a great assemblage, to welcome Henry V. after the battle of Agincourt, or more ominous and hostile demonstrations, as when Wat Tyler collected his followers there, or when Jack Cade, some seventy years later, did the same thing. A few fine old eighteenth-century houses still stand on the edge of the Heath, and an avenue, "Chesterfield Walk," perpetuates the name of one of the distinguished residents. Morden College, at the south-east corner of the Heath, is a fine old building of Wren's design, founded by Sir John Morden, for merchants trading with the East who, through unforeseen accidents, had lost their fortunes.

To the west of Blackheath there was once a Deptford

Common, but it has long since been built over, and, with the exception of the small Deptford Park, there is a large district of dense population without any open space. The nearest is Hilly Fields on the south. This is a steep, conical hill, with little beauty to recommend it, except its breezy height, and views over chimney-pots to the Crystal Palace. A large, bleak-looking building, with a small enclosure on the highest point—at present for sale—marked the West Kent Grammar School, does not improve the appearance of this open space. There are some 45 acres of turf, and a line of old elms and another of twisted thorns show that there were once hedgerows. There is some promiscuous planting of young trees, and iron railings, and of course a band-stand ; otherwise no particular " beautifying " has been attempted since it was opened to the public in 1896.

In the valley of the Ravensbourne, below the hill stretches the long, narrow strip of the Ladywell Re-creation Ground. It lies on either bank of the stream between Ladywell and Catford Bridge stations. It is intersected by railways, and the pathway passes some-times over, sometimes under the lines, and constant trains whizz by. But in spite of such drawbacks, the place has a special attraction in the stream which meanders through the patches of grass devoted to games. Where the stream has been untouched, and allowed to continue its course unmolested between iron railings, even the rail-ings cannot destroy a certain rural aspect it has retained. Alders and elms, with gnarled and twisted roots, lean over the banks, and hawthorns dip down towards the rather swiftly flowing water. When the land was bought for public use in 1889 the stream frequently overflowed its sandy banks, and one or two necessary cuttings were made across some of the sharpest curves, to allow a better flow of water. This has stopped all the objectionable flooding, but the melancholy part is that, having been obliged to make these imperative but necessarily artificial cuttings, the London County Council did not plant them with alders, thorns, and willows, like the pretty, natural stream; but instead, the islands thus formed, and the banks, were dotted about with box and aucuba bushes. The babbling stream seems to jeer at these poor sickly little black bushes, as if to say, " What is the good of bravely playing at being in the country, and trying to make believe trout may jump from my ripples and water-ousels pop in and out of my banks, if you dreadful Cockneys disfigure me like that?" Very likely it does not jar on the feelings of the inhabitants of Lewisham or Catford, but when public money is spent by way of improvement, it is cruel to mar and deform instead. Where the churchyard of St. Mary's, Lewisham, touches the stream is a pretty spot, but, in places, untidy little back-gardens are the only adornment; but that is not the fault of the London County Council.

Peckham Rye Common is more or less flat, without any special feature of interest, except at the southern end, which has been converted into a Park. The Rye— what a quaint name it is! and there is no very satisfactory derivation. It may either come from a stream of that name, long since disappeared, or from a Celtic word, *rhyn*, a projecting piece of land—Peckham Rye, the village on the spur of the hill, now known as Forest Hill and Honor Oak. This " Rye " has been a place of recreation from time immemorial, and at one time must have extended so as to embrace the smaller patches of common known as Nunhead Green (now black asphalt),

and Goose Green.   The Common was secured by purchase
from further encroachments in 1882.

The Park has much that savours of the country.   An
enclosure within it, is not open to the public, and for that
very reason is one of the most rural spots.   There is a
delightful public road across it, known as "the Avenue."
The old trees form an archway overhead, and on either
side of the fence the wood is like a covert somewhere
miles from London; brambles and fern and brushwood
make shelter for pheasants, and squirrels run up the
trees.   The farm-house, and its out-buildings with
their moss-grown tiled roofs, have nothing suburban
about them.   The front facing the Rye Common has a
notice to say it is the Friern Manor Dairy, but even that
is not aggressive, as the name carries back the history
to the time of Henry I., when the manor was granted to
the Earl of Gloucester, and on till it was given by his
descendants to the Priory of Halliwell, which held it
until the church property was taken by Henry VIII. and
granted to Robert Draper, and so on till modern days.
There is, besides this attractive farm, a regular piece of
laid-out garden, and a pond and well-planted flower-beds;
but the little walk among trees, beside a streamlet which
has been formed into small cascades, and crossed by rustic
bridges, is a more original conception, and is decidedly a
success, and a good imitation of a woodland scene.   The
contrast is all the greater as Peckham is so eminently
prosaic, busy, and unpicturesque; the old houses having
for the most part given place to modern suburban
edifices.

Due west of Peckham lies Clapham, the largest of
the South London Commons, 220 acres in extent;
although, being flat and compact in shape, it does not

appear larger than Tooting, which is really only 10 acres less, but of more rambling shape. The Common has suffered much less than most of its neighbours from enclosures. It was shared between two manors, Battersea and Clapham, and the rival lords and commonalities, each jealous of their own special rights, were more careful to prevent encroachments than was often the case. At one time Battersea went so far as to dig a great ditch to prevent the cattle of the Clapham people coming into its part of the ground. The other parish resisted and filled up the ditch, and was sued for trespass by Battersea, which, however, lost its case—this ended in 1718. The Common has an air of dignified respectability, and is still surrounded with some solid old-fashioned houses, although modern innovations have destroyed a great number of them. A nice old buttressed wall, over which ilex trees show their heads, and suggest possibilities of a shady lawn, carries one back to the time when Pepys retired to Clapham to "a very noble house and sweete place, where he enjoyed the fruite of his labour in great prosperity"; or to the days when Wilberforce lived there, and he, together with the other workers in the same cause, Clarkson, Granville Sharp, and Zachary Macaulay, used to meet at the house of John Thornton by the Common.

There is nothing wild now about the Common, and the numbers of paths which intersect it are edged by high iron railings, to prevent the entire wearing away of the grass. The beauty of the ground is its trees. They proclaim it to be an old and honoured open space, and not a modern creation. Only one tree has any pretentions to historical interest, having been planted by the eldest son of Captain Cook the explorer, but only a stump remains.

The ponds are the distinctive feature of the Common, and there are several of them dotted about, the joy of boys for bathing and boat-sailing. The origin of most of them has been gravel pits dug in early days. There is the Cock Pond near the church, the Long Pond, the Mount Pond, and the Eagle House Pond, some of them fairly large. The Mount Pond was at one time nearly lost to the Common, as about 1748 a Mr. Henton Brown, who had a house close by, and who kept a boat on the water, obtained leave to fence it in for his own private gratification. It was not until others followed Mr. Brown's example, and further encroachments began to frighten the parish, that it repented of having let in the thin end of the wedge. A committee was formed to watch over the interests of the Common lands, and took away Mr. Brown's privileges; but in spite of their vigilance other pieces were from time to time taken away. A little group of houses by the Windmill Inn are on the site of one of these shavings off the area, for a house called Windmill Place. The church was built on a corner of the Common in 1774, and has a peaceful, solid, dignified appearance, standing among fine old elms and away from the din of trams, which rush in all directions from the corner hard by. It was built to replace an older parish church, which was described as "a mean edifice, without a steeple" by a writer of the eighteenth century, who admired the "elegant" one which took its place. The present generation would hardly apply that epithet to the massive Georgian edifice, but it seems to suit its surroundings : substantial and unostentatious, recalling memories of the evangelical revival, it seems an essential part of the Common and its history.

Away to the south-west of Clapham lies Tooting

(why does the very name sound comic, and invariably produce a laugh ?), another Common, nearly as large, and much more wild and picturesque. Clapham is essentially a town open space, like an overgrown village green ; but on Tooting Common one can successfully play at being in the country. The trees are quite patriarchal, and have nothing suburban about them, except their blackened stems. There are good spreading oaks and grand old elms, gnarled thorns, tangles of brambles, and golden gorse. The grass grows long, with stretches of mossy turf, and has not the melancholy, down-trodden appearance of Clapham or Peckham Rye.

Fine elm avenues overshadow the main roads, and no stiff paths with iron rails, take away from the rural effect. Even the railway, which cuts across it in two directions, has only disfigured and not completely spoilt the park-like appearance. The disused gravel-pits, now filled with water, have been enlarged since the London County Council had possession ; and if only the banks could be left as wild and natural, as nature is willing to make them, they may be preserved from the inevitable stamp which marks every municipal park. The smaller holes, excavated by virtue of the former rights of digging gravel, and already overgrown, assist rather than take away from the charms of the Common.

Tooting Common consists of two parts, belonging to two ancient manors. The smallest is Tooting Graveney, which derives its name from the De Gravenelle family, who held the manor soon after the Conquest, on the payment of a rose yearly at the feast of St. John the Baptist. The larger half, Tooting Beck, takes its name from the Abbey of Bec in Normandy, which was in possession of the Manor from Domesday till 1414, when it came to the Crown. Both manors can be traced through successive owners until the rights were purchased in 1875 and 1873 by the Metropolitan Board of Works. The avenue of elms which runs right across the Common divides the two. Tooting Beck is more than twice the size of Graveney, and has the finest trees. One of the oldest elm trees, now encircled by a railing, was completely hollow, but now has a young poplar sprouting out of its shell. Tradition associates this tree particularly with Dr. Johnson, and though he did not compose his Dictionary under it, it is more than likely he often enjoyed the shade of what must have been a very old tree in his day. For fifteen years he was a constant visitor at Thrale Place close by. " He frequently resided here," says a contemporary guide-book, "and experienced that sincere respect to which his virtues and talents were entitled, and those soothing attentions which his ill-health and melancholy demanded." The house stood in 100 acres of ground between Tooting and Streatham Commons, and has since been pulled down and built over. During these years, no doubt, Tooting as well as Streatham Common was often trodden by the brilliant circle who drank tea and conversed with the accomplished Mrs. Thrale —Sir Joshua Reynolds, Burke, Garrick, Goldsmith—to all of them the woodland scenes of both Commons were familiar.

To prevent the too free use of the turf by riders, a special track has been made for them, skirting the Common, and passing down one of the finest avenues. It may save the grass from being too much cut up, but to those who don't feel called to gallop across the Common, the loss of the green sward under the tall feathery elms is a cause of regret. It is such, perhaps necessary, alterations which spoil the delusion of genuine country, otherwise so well counterfeited on Tooting Common. A charming time is when the may is out and the gorse ablaze with bloom, the chestnuts in blossom, and birds are singing all around; or if one happens to be there on a winter's day, when it is too cold for loungers or holiday-makers, there are moments when the nearness of streets and trams could be forgotten. The frosty air, and dew-drops on the vivid green grass, the brown of the fallen leaves, the dark stems clear against an amber sky, with the intense blue distance, which London atmosphere produces so readily, combine harmoniously into a telling picture, which remains photographed "upon that inward eye, which is the gift of solitude." The dream is as quickly dispelled. A sight, a sound, recalls the nearness of London, which makes its presence felt even when one is trying to play Hide-and-seek with the chimney-pots. How well Richard Jefferies, that inimitable writer on nature, describes his feelings in the neighbourhood of London, in spots only a little further from Hyde Park Corner than Tooting Beck :—

"Though my preconceived ideas were overthrown by the presence of so much that was beautiful and interesting close to London, yet in course of time I came to understand what was at first a dim sense of something wanting. In the shadiest lane, in the still pine-woods, on the hills of purple heather, after brief contemplation there arose a restlessness, a feeling that it was essential to be moving. In no grassy mead was there a nook where I could stretch myself in slumbrous ease and watch the swallows ever wheeling, wheeling in the sky. This was the unseen influence of mighty London. The strong life of the vast city magnetised me, and I felt it under the calm oaks."

The most remote of London open spaces in this direction is Streatham, to the south-east of Tooting, close to Norwood, and on the very extremity of the County of London. Much smaller than the other commons, it possesses attractions of its own. It is less spoilt by modern buildings than any of these once country villages, but ominous boards foretell the rapid advance of the red-brick villa. The houses which now overlook the upper part are substantial, in the solid, simple style of the eighteenth century. In those days Streatham possessed a mineral spring, and for a few years people flocked to drink at it. But long before the end of the eighteenth century other more fashionable watering-places had supplanted it, and in 1792 Streatham is described as " once frequented for its medicinal waters." The spring was in the grounds afterwards belonging to a house called the Rookery, and near the house called Wellfield, on the southern side of the Common. The waters were said to be so strong that three glasses of Streatham were equivalent to nine of Epsom. Although so near London, the journey to the springs presented some dangers, as this was one of the most noted localities for footpads and highwaymen. The woods of Norwood, which came close to the Common, afforded covert and an easy means of escape. This road from London, which went on to Croydon and Brighton, had such a bad reputation that the risk of an adventure must have counterbalanced some of the health-giving properties to any nervous invalid ! The lower part of the Common, near the road, is flat and open, and not particularly inviting.

The charms of the top of the hill are all the more delightful, as they come as a surprise. There are fine old trees, and a wealth of fern, thorns, and bramble, and the short grass is exchanged for springy turf the moment the crest of the steep hill is reached. But by far the greatest surprise is the glorious view. Away and away over soft, hazy, blue country the eye can reach. It may or may not be true that Woolwich, Windsor, and Stanmore can be seen : nobody will care who gazes over that wide stretch of country bathed in a mysterious light, perhaps with the rays of the sun, like golden pathways from heaven, carrying the thoughts far from the prosaic villas or harrowing slums concealed at one's feet. Only the wide expanse and the waving bracken and tangled brushwood fill the picture —while one rejoices that such a beautiful scene should be within the reach of so many of London's toilers.

Wandsworth is among the least beautiful and the most cut-up of the commons. Large and straggling in extent, it has been so much encroached upon that roads, and houses, and railways cross it. It is narrowed to a strip in places, and all the wildness and all the old trees have gone. Some young avenues by the main road have been planted, and no more curtailments can be perpetrated, as it was acquired for the use of the public in 1871. For many years the encroachments had roused the inhabitants, and about 1760 a species of club was formed to protect the rights of the commoners. When enclosures took place, the members all subscribed and went to law, and often won their cases. The head was called the " Mayor of Garratt," from Garratt Lane, near the Common, where a "ridiculous mock election " was held. A mob collected, and en-

couraged by Foote, Wilkes, and others, witty speeches
were made. Foote wrote a farce called "The Mayor
of Garratt," which for some time gave the ceremony
no small celebrity. The rowdyism becoming serious
at the sham elections, they were suppressed in 1796.
When the Common was eventually saved, it was in a
bad and untidy state : quantities of gravel had been
dug, and holes, some of them filled with water, were
a danger ; the trees had all disappeared, and the whole
surface was bare and muddy. It has improved since
then, but there is nothing picturesque left. The "Three
Island Pond," which is supposed to be its greatest
beauty, is stiff, formal, and new-looking, with a few
straggly trees growing up. Still it is safely preserved
as an open space, and makes a good recreation ground.

All round London, besides the larger commons,
smaller greens are to be found, which are survivals of
the old village greens. They recall the time when
London was a walled city, and thickly scattered round
it were the little hamlets which have now been absorbed
by the ever-growing, monster town.

There is little that is distinctive about them. For
the most part they are simply open spaces of well-worn
turf without trees. Shepherd's Bush is one of these.
Brook Green, in Hammersmith, not very far from it,
has the remains of a few fine elm trees. In Fulham
there are Parson's Green and Eel Brook Common.
Away in South London, Goose Green and Nunhead
Green are other examples where grass is even more
inconspicuous.

On the north lies Paddington Green, which is small
in extent, but close to the large graveyard turned into a
public garden. In the centre of the Green a statue to

Victoria Manser
1906

STATUE OF MRS. SIDDONS, PADDINGTON GREEN

Mrs. Siddons, by Chevaliand, was erected in 1897, as she lived in the neighbourhood when Paddington was still rural. There is nothing beautiful about the asphalt paths between high iron railings surrounding the small space of grass and trees. Some of the other greens are more of the ordinary public garden type. Islington Green has been planted with trees, and outside the railings stands a statue of Sir Hugh Myddelton, who died in 1631, representing him holding a plan of the New River. Stepney was once a very large green, and has still $3\frac{1}{4}$ acres of garden cut up into four sections. Some quaint old houses, wood with tiled roofs, and good seventeenth-century brick ones, still overlook the gardens. The gardens have been made exactly like every other, with a slightly serpentine path, a border running parallel in irregular curves not following the line of the path, and trees dotted about. One really fine, thick-stemmed laburnum shows how well that tree will do in smoke, and some curious old wooden water-pipes dug up in 1890, dating from 1570, are placed at intervals in the grass.

Camberwell has one of the large village greens of South London, and has been made into a satisfactory garden. All the trams seem to meet there, but in spite of the din it is a pleasant garden in which to rest. The $2\frac{1}{2}$ acres are well laid out, and the clipped lime-trees round the railings are a protection from the street which other places would do well to copy. When the trees are in leaf the garden is partially hidden even from those on the tops of omnibuses.

These greens scattered round London help to connect the larger areas, thus forming links in the chain of open spaces which encircles London. These natural recreation grounds are the admiration of all foreigners, and a priceless boon to the citizens, ensuring the preservation of green grass and green trees to refresh their fog-dimmed eyes, at no great distance from the throng of city life.

CHAPTER IX

SQUARES

Fountains and Trees our wearied Pride do please,
Even in the midst of Gilded Palaces ;
And in our Towns, that Prospect gives Delight,
Which opens round the Country to our Sight.

—Lines in a Letter from Sprat to Sir Christopher
Wren on the Translation of Horace.

OTHING is more essentially cha-
racteristic of London than its
squares. They have no exact
counterpart in any foreign city.
The iron railings, the enclosure
of dusty bushes and lofty trees,
with wood-pigeons and twittering
sparrows, have little in common
with, say for instance, the Place Vendôme in Paris,
or the Grand' Place in Brussels, or Madison Square,
New York. The vicissitudes of some of the London
Squares would fill a volume, but most of them have
had much the same origin. They have been built
with residential houses surrounding them, and though
some have changed to shops, and in others the houses
are dilapidated and forsaken by the wealthier classes,
nearly every one has had its day of popularity.

In some of those now deserted by the world
of fashion, the gardens have been opened to the

public, but by far the greater number of squares are maintained by the residents in their neighbourhood, who have keys to the gardens. But even though they are kept outside the railings the rest of the public receive a benefit from these air spaces and oxygen-exhaling trees. Sometimes the public get more direct advantage, as in such cases as Eaton Square, where seats are placed down the centre on the pavement under the shade of the trees inside the rails, and are much frequented in hot weather; or in Lower Grosvenor Gardens, which are open for six weeks in the autumn, when most of the residents in the houses are absent.

Squares are dotted about nearly all over London, but they can, for the most part, be grouped together. There are the older ones, of different sizes, and varying in their modern conditions. Among such are Lincoln's Inn Fields, Charterhouse, Soho, Golden, Leicester, and St. James's Squares. Then there is the large Bloomsbury group, and further westward the chain of squares begins with Cavendish, Manchester, Portman, on the north, and Hanover and Grosvenor to the south of Oxford Street. Then follow the later continuations of the sequence—Bryanston, Montagu, and so on to Ladbroke Square, nearly to Shepherd's Bush. To the south of the Park lies the Belgravia group, with more and more modern additions stretching westward till they join the old village of Kensington, with dignified squares of its own, or till their further multiplication is checked by the River.

To describe most of these squares would imply a vast amount of vain repetition. Few have anything original either in design or planting. The majority have elms and planes mixed with ailanthus, while aucubas, euonymous, and straggling privet form the staple product of the encircling borders, with a pleasant admixture of lilac and laburnum, and generally a good supply of iris facing the gravel pathway. A few annuals and bedding-out plants brighten the borders in summer, and some can boast of one or two ferns. Occasionally the luxury of a summer-house is indulged in, and here and there a weeping ash has been ventured upon by way of shelter; a secluded walk or seat is practically unknown. The older gardens have some large trees, and the turf in all of them is good, and when it is with "daisies pied" it forms the chief delight of the children who play there. It may be that the distance of Notting Hill Gate from the smoke of the East End has encouraged more enterprise in gardening; certainly the result of the planting in Ladbroke Square is satisfactory. Several healthy young oaks are growing up; and a fountain and small piece of formal gardening round it, on the highest point of the long, sloping lawn, is effective. In the older squares, such as Grosvenor Square, the bushes are high, and the openings so well arranged that the lawns in the centre are perfectly private, and hidden from the streets. In the less ancient ones, such as Eccleston and Warwick, Connaught and Montagu Squares, the long, narrow strip leaves little scope for variation.

An innovation of the usual square is to be seen in Duke Street, Grosvenor Square. This small square, which was laid out as a garden with sheltered seats, was made when the new red-brick dwellings replaced the smaller and more crowded houses. The middle is now the distributing centre of an electric power-station, but the roof is low and flat, and has been successfully transformed into a formal garden, with trees in tubs and boxes of flowers.

Some of the squares have finer trees than others, and in many a statue is a feature. Originally these statues formed the central object towards which the garden paths converged, but most of the central statues have

WINTER GARDEN, DUKE STREET, GROSVENOR SQUARE

been moved, though in a few, like St. James's and Golden Squares, they are still in the middle. These statues were evidently a good deal thought of by Londoners, but they did not strike foreigners as very good. In one of Mirabeau's letters he writes in 1784 from London: "The public monuments in honour of Sovereigns, reflect little honour on English Sculpture. . . . The Statues of the last Kings, which adorn the

STATUE OF PITT, HANOVER SQUARE

STATUE OF LITT, PARADISE SQUARE.

Squares in the new quarters of London, being cast in
brass or copper, have nothing remarkable in them but
their lustre; they are doubtless kept in repair, cleaned
and rubbed with as much care as the larger knockers at
gentlemen's doors, which are of similar metal." The
usual plan now is to place the statue facing the street,
where a background of green shows it off to the
passer-by. Thus Lord George Bentinck is prominent
in Cavendish Square, from which the equestrian central
statue of the Duke of Cumberland has gone; and from
Hanover Square, built about the same time as Cavendish
(between 1717–20), Chantrey's statue of Pitt gazes
down towards St. George's Church. In Grosvenor
Square no statue has replaced the central one of
George I. by Von Nost, which was placed there in 1726,
and is described by Maitland as a " stately gilt equestrian
statue." This Square is older than the two last mentioned, having been built in 1695. In those days each
of the spacious houses had its large garden at the back,
with a view of the country away to Hampstead and
Highgate. The garden was designed by Kent, but a plan
of it about 1750 shows a considerable difference between
the arrangements then and now, although some details
are the same. The raised square of grass in the centre
where the statue stood has now a large, octagonal, covered
seat, apparently formed with the old pedestal. The
walk round and the four wide paths to the centre are
retained, but the smaller intersecting paths are replaced
by lawns on which grow some fine old elms. The railings with stone piers and handsome gates, shown in the
engraving, have given place to much less ornamental
iron rails.

Manchester Square is of later date. It was an open

space approached by shady lanes from Cavendish Square
for some fifty years after that was built.   The houses in
Manchester Square were not begun till 1776—some ten
years after the commencement of Portman Square.   This
district was all very semi-rural and unfinished until much
later.   Southey, in a letter, writes of Portman Square as
" on the outskirts of the town," and approached " on
one side by a road, unlit, unpaved, and inaccessible by
carriages."   The large corner house, now occupied by
Lord Portman, was built for Mrs. Montagu, " Queen
of the Blue Stockings," and during her time " Montagu
House " was the salon to which the literary celebrities
of the day flocked.   When Mrs. Montagu moved
there from Hill Street she wrote to a friend, " My
health has not been interrupted by the bad weather
we have had ; I believe Portman Square is the Mont-
pellier of England."   In the centre of the Square garden
was planted a " wilderness," after the fashion of the day,
and early in the nineteenth century, when the Turkish
Ambassador resided in the Square, he erected a kiosk in
this " wilderness," where he used to smoke and imagine
himself in a perfumed garden of the East.   It is still
one of the best kept-up of the squares.

Berkeley Square dates from nearly the same time as
Grosvenor, having been begun in 1698, on the site of the
extensive gardens of Berkeley House, which John Evelyn
so much admired, and where flourished the holly hedges
of which he advised the planting.   The central statue
here was one by Beaupré and Wilton of George III.,
which was removed in 1827, and the base of the statue
made into a summer-house.   In the place of the usual
statesman, a drinking fountain, with a figure pouring
the water—the gift of the Marquess of Lansdowne—

has been placed outside the rails at the southern end. The plane trees are very fine, and were planted at the end of the eighteenth century, it is said, by Mr. Edward Bouverie in 1789. The plane has been so long grown in London these cannot be said with certainty to be the oldest, as is so often stated. Some in Lincoln's Inn Fields are decidedly larger. In 1722 Fairchild writes in praise of the plane trees, about 40 feet high, in the churchyard of St. Dunstan-in-the-East. Loudon mentions one at the Physic Garden, planted by Philip Miller, which was 115 feet high in 1837 (a western Plane—not the great oriental Plane which fell down a few years ago). The western Plane (*Platinus occidentalis*) was introduced to this country many years after the eastern Plane (*Platinus orientalis*). The tree most common in town is a variety of eastern Plane called *accrifolia*, known as the "London Plane": this must have been a good deal planted all through the eighteenth century, so it is difficult to assign to any actual tree the priority.

St. James's Square is older than any of the squares already glanced at, having been built in the time of Charles II. It was known as Pall Mall Field or Close, originally part of St. James's Fields, and the actual site of the Square was a meadow used by those attached to the Court as a sort of recreation ground. Henry Jermyn, Earl of St. Albans, leased it in 1665 from Charles II., and began to plan the Square or "Piazza," as it was called at first. The deadly year of the Plague, followed by the Great Fire, delayed the building, and the houses were not finished and lived in till 1676. No. 6 in the Square, belonging to the Marquess of Bristol, has been in his family since that time. Every one of the fine old houses has its story of history and romance. Here Charles II. was frequently seen visiting Moll Davis, Sir Cyril Wyche, and the Earl of Ranelagh. The Earl of Romney, and the Duke of Ormond, and Count Tallard the French Ambassador, are names connected with the Square in William III.'s time, and Josiah Wedgwood lived at No. 7. But these and many other historical personages did not look from their windows on to a well-ordered garden, and the Court beauties did not wander with their admirers under the spreading trees. The centre of the Square was left open, and merely like a field. The chief use to which the space seems to have been put was for displays of fireworks. One of the great occasions for these was after the Peace of Ryswick, but unfortunately they were not always very successful. An eye-witness, writing to Sir Christopher Hatton, says of Sir Martin Beckman, who had the management of them, that he "hath got the curses of a good many and the praises of nobody." The open space eventually became so untidy that the residents in 1726 petitioned Parliament to allow them to levy a special rate to "cleanse, adorn, and beautify the Square," as "the ground hath for some years past lain, and doth now lie, rude and in great disorder, contrary to the design of King Charles II., who granted the soil for erecting capital buildings." So badly used was it that even a coach-builder had erected a shed in the middle of it, in which to store his timber. Strong measures were taken, and any one "annoying the Square" after May 1, 1726, was to be fined 20s., and any one encroaching on it, £50. No hackney coach was allowed to ply there, and unless a coachman, after setting down his fare, immediately drove out of the Square, he was to be fined 10s. The whole place was levelled and paved, and a round basin of water, which was intended to have a fountain in it, and never did, was dug in the centre. Round it ran an octagon railing with stone obelisks, surmounted with lamps at each angle. A road of flat paving-stones with posts went round the Square in front of the houses ; the rest was paved with cobble stones. As early as 1697 it was proposed to place a statue of William III., and figures emblematical of his victories, in the Square, but nothing was done. In 1721 the Chevalier de David tried to get up a subscription for a sum of £2500 for a statute of George I. to be done by himself and set up, but, as he only collected £100 towards it, that scheme also fell through. Once more an effort was made which bore tardy fruit, for in 1724 Samuel Travers bequeathed a sum of money by will " to purchase and erect an equestrian statue in brass to the glorious memory of my master, King William III." Somehow this was not carried out at the time, but in 1806 the money appeared in a list of unclaimed dividends, and John Bacon the younger was given the commission to model the statue, which was cast in bronze at the artist's own studio in Newman Street, and put up in the centre of the pond. Thus it remained until towards the middle of last century the stagnant pool was drained. In the 1780 riots the mob carried off the keys of Newgate and flung them into this basin, where years afterwards they were found. It was 150 feet in diameter, and 6 or 7 feet deep. When the pond was drained, the garden was planted in the form it now is, and the statue left standing in the centre. St. James's is still one of the finest residential squares in London, and the old rhyme, picturing the attractions in store for the lady of quality who became a duchess and lived in the Square.

P

might have been written in the twentieth instead of the eighteenth century.

> " She shall have all that's fine and fair,
> And the best of silk and satin shall wear ;
> And ride in a coach to take the air,
> And have a house in St. James's Square."

Less cheerful has been the fate of Golden Square, which has a forsaken look, and the days when it may have justified its name are past.   Originally Gelding Square, from the name of an inn hard by, the grander-sounding and more attractive corruption supplanted the older name.   Another derivation for the word is also given—" Golding," from the name of the first builder ; but anyhow it was called Golden Square soon after it came into being.   The houses round it were built about the opening years of the eighteenth century, when the dismal memories of the Plague were growing faint.   For the site of Golden Square, "far from the haunts of men," was one of the spots where, during the Plague, thousands of dead were cast, by scores every night.   These gloomy scenes forgotten, the Square was built, and at one time fashionable Lord Boling-broke lived here, while Secretary for War.   It is still "not exactly in anybody's way, to or from anywhere." The garden is neat, with a row of trees round the Square enclosure, and a path following the same lines. In the centre stands a statue of George II., looking thoroughly out of place, like a dilapidated Roman emperor.   It was bought from Canons, the Duke of Chandos's house, near Edgware, when the house was pulled down and everything sold in 1747.   There are a few seats, but they are rarely used, and it has a

STATUE OF WILLIAM III. IN ST. JAMES'S SQUARE

very quiet and dreary aspect when compared with the
cheerful crowds enjoying the gardens in its larger
neighbour, Leicester Square. This was known as
Leicester Fields, and was traversed by two rows of
elm trees; and even after the houses round it were
begun, about 1635, the name of Fields clung to
it. The ground was part of the Lammas Lands belonging to the parish of St. Martin's-in-the-Fields, and
Robert Sydney, Earl of Leicester, who built the house
from which the Square takes its name, paid compensation for the land, to the poor of the parish £3 yearly.
The house occupied the north-east corner of the Square,
and in after years became famous as a royal residence.
It has been called "the pouting-place" of princes,
as it was to Leicester House that the Prince of Wales
retired when he quarrelled with his father, George I.;
and there Caroline the Illustrious gathered all the
dissatisfied courtiers, and such wit and beauty as could
be found, round her. When he became George II.,
and quarrelled in his turn with his son, Frederick,
Prince of Wales, the latter came to live in Leicester
House. The statue of George I. which stood in the
centre of the garden was, it was said, put up by
Frederick, with the express purpose of annoying his
father. A view of the Square in 1700, shows a neatly-
kept square garden with four straight walks, and trees
at even distances, and Leicester House standing back,
with a fore-court and large entrance gates, and a garden
of its own with lawns and statues at the back. Savile
House, next door to Leicester House, on the site
of the present Empire Theatre, was also the scene
of many interesting incidents, until it was practically
destroyed during the Gordon Riots. The list of great

names connected with the Square is too long to recite, but four of the greatest are commemorated by the four busts in the modern garden — Sir Joshua Reynolds, Hogarth, John Hunter, the eminent surgeon, and Sir Isaac Newton. But before these monuments were erected Leicester Square Garden had gone through a period of decay. It was left unkept up and uncared for; the gilt statue was tumbling to pieces, and was only propped up with wooden posts. The garden from 1851 for ten years, was used to exhibit the Great Globe of Wylde, the geographer, who leased the space from the Tulk family, then the owners of the land. Leicester House, after it ceased to be a royal residence, was in the hands of Sir Ashton Lever, who turned it into a museum, which was open from 1771 to 1784, but failed to obtain much popularity. The collection was dispersed, and soon after the house was pulled down and the site built over, and the Square was allowed to get more and more untidy. Several efforts were made to purchase it for the public, but the price asked was prohibitive, as the owners wished to build on it. When, however, after much litigation, the Court of Appeal decided it could not be built on, but must be maintained as an open space, they were more ready to come to terms. A generous purchaser came forward, Mr. Albert (afterwards Baron) Grant, who bought the land, laid it out as a garden, and presented it to the public, to be kept up by the Metropolitan Board of Works. The plans for the newly-restored garden, were made for Mr. Grant by Mr. James Knowles, and the planting done by Mr. John Gibson, who was then occupied with the sub-tropical garden in Battersea Park. The statue of

Shakespeare in the centre, and the four busts, were also the gift of the same public benefactor, who presented the Square complete, with trees, statues, railings, and seats, in 1874.

Soho Square was another of the fashionable squares of London, now gloomy and deserted by its former aristocratic residents. The gardens are kept up for the benefit of those living in the Square only, and are not enjoyed by the masses, like Leicester Square. Maitland describes the building and consecration of St. Anne's, Soho, or, as he calls it, St. Anne's, Westminster, which was in 1685 separated from St. Martin's-in-the-Fields, and a new parish created, just in the same way as scores of parishes have to be treated nowadays, to meet the needs of the much more rapidly-growing population. Of the new parish, he says the only remarkable things were " its beautiful streets, spacious and handsome Church, and stately Quadrate, denominated King's-Square, but vulgarly Soho-Square." Various suggestions have been made as to the origin of the name, and the most popular explanation is that it was a hunting-cry used in hunting hares, which sport was indulged in over these fields. The word Soho occurs in the parish registers as early as 1632. When first built the Square was called King Square, from Geoffrey King, who surveyed it, not after King Charles II. But the old name of the fields became for ever attached to the Square, to the entire exclusion of the more modern one, after the battle of Sedgemoor. Monmouth's supporters on that occasion took the word Soho for their watchword, from the fact that Monmouth lived in the Square. In 1690 John Evelyn notes that he went with his family " to winter at Soho in the Great Square." Monmouth House was built by Wren, when the Square was begun in 1681, and it was pulled down, to make room for smaller houses on the south side of the Square, in 1773. There are some fine old trees in the garden, and a statue of Charles II. used, till the middle of last century, like the one in St. James's Square, to stand in a basin of water, with figures round it, emblematic of the rivers Thames, Severn, Tyne, and Humber, spouting water. Nollekens, the sculptor, who was born in 28 Dean Street, Soho, in 1738, recalled how he stood as a boy " for hours together to see the water run out of the jug of the old river-gods in the basin in the middle of the Square, but the water never would run out of their jugs but when the windmill was going round at the top of Rathbone Place." The centre of the Square was in 1748 " new made and inclosed with iron railings on a stone kirb," and "eight lamp Irons 3 ft. 6 in. high above the spikes in each of the Eight corner Angles " : the " Channell all round the Square" was paved with " good new Kentish Ragg stones."

Beyond Oxford Street are collected a great number of squares in the district of Bloomsbury. They are all surrounded by solid, well-built houses, which seem to hold their own with dignity, even though fashion has moved away from them westward. Before the squares arose, this was the site of two great palaces with their gardens. One of them, Southampton House, afterwards known as Bedford or Russell House, was where Bloomsbury Square now is. In 1665, February 9, Evelyn notes that he "dined at my Lord Treasurer's the Earl of Southampton, in Bloomsbury, where he was building a noble square or piazza, a little town ; his own house stands too low— some noble rooms, a pretty cedar chapel, a naked garden to the North, but good air." This house was pulled down in 1800, and Russell Square was built on the garden. Both Bloomsbury, or Southampton Square, as it was sometimes called, and Russell Square have good trees, and in each garden there is a statue by Westmacott. Charles James Fox, seated in classical drapery, erected in 1816, looks down Bedford Place, where stood Southampton House, towards the larger statue, with elaborate pedestal and cupids, of Francis, Duke of Bedford, in Russell Square. This is one of London's largest Squares, being only about 140 feet smaller than Lincoln's Inn Fields, and included most of the garden of Southampton House, with its fine limes, and a large locust-tree, *Robinia pseudo acacia*.

The laying out is more original in design than most of the squares, having been done by Repton in 1810. In Repton's book on Landscape Gardening he goes fully into his reasons for the design of Russell Square. "The ground," he said, "had all been brought to one level plain at too great expense to admit of its being altered." He approves of the novel plan of placing the statue at the edge instead of in the usual position in the centre of the Square. "To screen the broad gravel-walk from the street, a compact hedge is intended to be kept clipt to about six feet high; this, composed of privet and hornbeam, will become almost as impervious as a hedge of laurels, or other evergreens, which will not succeed in a London atmosphere." He says he has not "clothed the lawn" with plantation, so that children playing there could be seen from the windows, to meet "the particular wishes of some mothers." "The outline of this area is formed by a walk under two rows of lime-trees, regularly planted at equal distances, not in a perfect circle, but finishing towards the statue in two straight lines." He imagines that fanciful advocates of landscape gardening will object to this as too formal, and be "further shocked " by learning that he hoped they would be kept cut and trimmed. Within were to be "groves in one quarter of the area, the other three enriched with flowers and shrubs, each disposed in a different manner, to indulge the various tastes for regular or irregular gardens." He ends his description by saying : " A few years hence, when the present patches of shrubs shall have become thickets —when the present meagre rows of trees shall have become an umbrageous avenue—and the children now in their nurses' arms shall have become the parents or grand-sires of future generations—this square may serve to record, that the Art of Landscape Gardening in the beginning of the nineteenth century was not directed by whim or caprice, but founded on a due consideration of utility as well as beauty, without a bigoted adherence to forms and lines, whether straight, or crooked, or serpentine."

Repton always put forth his ideas in high-sounding language, often not so well justified as in the present case. The lime-trees have been allowed to grow taller than he desired, and yet are not fine trees from having at one period been kept trimmed ; but they certainly form an attractive addition to the usual design, and looking at them, after nearly a hundred years, from the outside, where they form a background to the statue, the effect in summer is very attractive.

Bedford Square is on the gardens of the other great house—Montagu House, built by the Duke of Montagu. Evelyn also notes going to see that. In 1676, "I dined," he says, "with Mr. Charleton and went to see Mr. Montagu's new palace near Bloomsbury, built by

Mr. Hooke of our Society [the Royal] after the French manner." This house was burnt down ten years later, and rebuilt with equal magnificence ; but when the Duke moved to Montagu House, Whitehall, in 1757, it became the home of the British Museum. The old house was pulled down and the present building erected in 1845. The Square was laid out at the end of the eighteenth century on the gardens and the open fields of the parish of St. Giles-in-the-Fields beyond. Lord Loughborough lived in No. 6, and after him Lord Eldon from 1804 to 1815. At the time of the Gordon Riots in 1780, when Lord Mansfield's house was plundered, troops were stationed near, and a camp formed in the garden of the British Museum. That garden was also of use when, in March 1815, Lord Eldon's house in Bedford Square was attacked by a mob, and he was forced to make his escape out of the back into the Museum garden.

Of Queen's Square, built in Queen Anne's time, but containing a statue of Queen Charlotte, and all the other squares of this district there is little of special interest to record directly connected with their gardens. They all have good trees, and are kept up much in the same style.

Red Lion Square is an exception. It has a longer history, and now its garden differs from the rest, as it is open to the public, and a great boon in this crowded district. It takes its name from a Red Lion Inn, which stood in the fields long before any other houses had grown up near it. It was to this inn that the bodies of the regicides Cromwell, Ireton, and Bradshaw were carried, when they were exhumed from Westminster Abbey and taken, with all the horrible indignities meted out to traitors, to Tyburn. A tradition, probably without foundation, was for long current that a rough stone obelisk, which stood afterwards in the Square, marked the spot where Cromwell's body was buried by friends who rescued the remains from the scaffold. The houses were built round it at the end of the seventeenth century, but the space in the middle seems, like all other squares at this time, to have been more or less a rubbish heap, and a resort of "vagabonds and other disorderly persons." In 1737 the inhabitants got an Act of Parliament to allow them to levy a rate to keep the Square in order. A contemporary, in praising this determination to beautify the Square, "which had run much to decay," hopes that "Leicester Fields and Golden Square will soon follow these good examples." The "beautifying" consisted in setting up a railing round it, with watch-houses at the corners, while the obelisk rose in the centre out of the rank grass.

The present garden, when first opened to the public, was managed by the Metropolitan Gardens Association, but since 1895 the London County Council have looked after it; the inhabitants having made a practically free gift of it for the public benefit. The nice old trees, flowers, seats, and fountain make it a much less gloomy spot than during any time of its history since the Red Lion kept solitary watch in the fields.

The largest of all the squares is Lincoln's Inn Fields. The garden, which is $7\frac{1}{4}$ acres in extent, was, after many lengthy negotiations, finally opened to the public in 1895. The fine old houses which survive, show the importance and size of Inigo Jones's original conception. It has been said that the Square is exactly the same size as the base of the Great Pyramid, but this is not the case. The west side, which was completed by Inigo Jones, was begun in 1618, but the centre of the Square was left an open waste till long after that date. The Fields, before the building commenced, were used as a place of execution, and Babington and his associates met a traitor's death, in 1586, on the spot where it was supposed they had planned some of their conspiracy. The surrounding houses had been built, and the ground was no longer an open field when William, Lord Russell, was beheaded there in 1683. The scaffold was erected in what is now the centre of the garden. The Fields for many years bore a bad name, and were the haunt of thieves and ruffians of all sorts. When things reached such a climax, that the Master of the Rolls was knocked down in crossing the Fields, the centre was railed in. This was done about 1735, with a view to improving their condition, and they remained closed, and kept up by the inhabitants, until a few years ago. The chief feature in the pleasant gardens now are the very fine trees. There are some patriarchal planes, with immense branches, under which numbers of people are always to be seen resting. The houses, Old Lindsay House, Newcastle House, the College of Surgeons, Sir John Soane's Museum, with long histories of their own, and all the lesser ones, with a sleepy air of dingy respectability and ancient splendour, now look down on a most peaceful, well-kept garden, and Gay's lines of warning are no longer a necessary caution :—

> "Where Lincoln's Inn wide space is rail'd around,
> Cross not with venturous step ; there oft is found
> The lurking thief, who, while the daylight shone,
> Made the walls echo with his begging tone ;
> That crutch, which late compassion moved, shall wound
> Thy bleeding head, and fell thee to the ground."

Adjoining the Fields is New Square, which used to be known as Little Lincoln's Inn Fields, and earlier still as

Fickett's Field or Croft. It was built in 1687. Fickett's
Fields occupied a wider area, and until 1620 they, like
the larger Fields, were a place of execution. The site of
New Square was planted and laid out in very early days.
The Knights of St. John in 1376 made it into a walking
place, planted with trees, for the clerks, apprentices, and
students of the law. In 1399 a certain Roger Legit was
fined and imprisoned for setting mantraps with a "mali-
cious intention to maim the said clerks and others," as
they strolled in their shaded walks. This Square, like all
others, went through phases of being unkept and untidy,
but was finally remodelled, into its present neat form, in
1845.

Eastwards, into the heart of London there are the
squares which are the remains of the open ground
without the City walls. Charterhouse Square, which is
now a retired, quiet spot with old houses telling of a
former prosperity, has a history reaching back to the
fourteenth century. In the days of the Black Death,
when people were dying so fast that the Chronicler of
London, Stowe, says that "scarce the tenth person of all
sorts was left alive," the "churchyards were not sufficient
to receive the dead, but men were forced to chuse out
certaine fields for burials: whereupon Ralph Stratford,
Bishop of London, in the yeere 1348 bought a piece of
ground, called *No man's land*, which he inclosed with a
wall of Bricke, and dedicated for buriall of the dead,
builded thereupon a proper Chappell, which is now
enlarged, and made a dwelling-house: and this burying
plot is become a faire Garden, retaining the old name
of Pardon Churchyard." It was very soon after this
purchase, that the Carthusian monastery was founded
hard by; but although the land was bought by the

Order, Pardon Churchyard was maintained as a burial-ground for felons and suicides. After the dissolution of the monasteries, when Charterhouse School and Hospital had been established by Thomas Sutton, the houses round the other three sides of the Square began to be built. One of the finest was Rutland House, once the residence of the Venetian Ambassador. It is still a quiet, quaint place of old memories; and the garden, with two walks crossing each other diagonally, and some fair-sized trees, has a solemn look, as if, even after all the centuries that have passed, it had some trace of its origin. Finsbury Circus and Finsbury Square are very different. They are more modern, bustling places which have entirely effaced the past. That they were, for long years, the most resorted to of open spaces, where Londoners took their walks is well-nigh forgotten, except in the name Finsbury, or Fensbury, the fen or moor-like fields without the walls. Bethlehem Hospital, known as Bedlam, was, for many generations, the only large building on the Fields. Finsbury Square was begun more than a hundred years ago, and but for the few green trees, nothing suggests the former country origin. Trinity Square, by the Tower, is so unique in aspect and association that it must be mentioned. In the sixteenth century the " tenements and garden plots " encroached on Tower Hill right up to the " Tower Ditch," and from the earliest time some kind of garden existed at the Tower. When it was a royal residence, frequent entries appear in the accounts of payments for the upkeep of the garden. Although so much has changed, and the wild animals that afforded amusement for centuries are removed, it is pleasant to see the moat turned into walks, and well planted with iris and hardy plants, and making quite a bright show in summer, in contrast to the sombre grey walls.

Away in the East End there are numbers of other gloomy little squares whose gardens are the playground of the neighbourhood. They are useful spaces of air and light, and the few trees and low houses surrounding them give a little ventilation in some of the very crowded districts. They are all much alike; in some more care has been taken in the planting and selection of the trees than in others. There is De Beauvoir Square, Dalston; Arbour Square, off the Commercial Road; York Square, Stepney; Wellclose, near the Mint and London Docks; Trafalgar Square, Mile End; and many others dotted about among the dismal streets. Turning to the West End again, the largest of the square spaces is Vincent Square, which forms the playground of the Westminster boys. It derives its name from Dr. Vincent, the head-master who was chiefly instrumental in obtaining it for the use of the boys. It was first marked out in 1810, and enclosed by railings in 1842. The 10 acres of ground were part of Tothill Fields, and the site was a burial-place in the time of the Great Plague.

There is nothing of historical interest in the Squares of Belgravia. The ground covered by Belgrave Square was known as Five Fields, which were so swampy that no one had attempted to build on them. It was the celebrated builder, Thomas Cubitt, who in 1825 was able by draining, and removing clay, which he used for bricks, to reach a solid foundation, and in a few years had built Belgrave and Eaton Squares and the streets adjoining. The site of the centre of Belgrave Square was then a market-garden. Ebury Square, the garden of which is open to the public, and tastefully laid out,

was built about 1820. The farm on that spot, which
in 1676 came to the Grosvenor family, was a farm of
430 acres in Queen Elizabeth's time, and is mentioned
as early as 1307, when Edward I. gave John de
Benstede permission to fortify it. There was only one
road across the swampy ground from St. James's to
Chelsea, and that was the King's Road, which followed
the line of the centre of Eaton Square. There were,
however, numerous footpaths, infested by footpads and
robbers at night, and bright with wild flowers and
scented by briar roses by day. There is a great sameness among all the squares between Vauxhall Bridge
and the Pimlico Road. Of this latter original-sounding
name there seems no satisfactory explanation. The
space between Warwick Street and the river, was in
old times occupied by the Manor House of Neyte,
and in later days by nurseries and a tea garden, known
as the Neat House. The ground near Eccleston Square
was an osier bed. The whole surface was raised by
Cubitt, with soil from St. Katherine's Docks in 1827,
and the houses built, and square gardens laid out; Eccleston in 1835, Warwick 1843, St George's 1850, and so
on until the whole was covered. The gardens are all
in the same style, and have no horticultural interest.
The garden in front of Cadogan Place varied most from
the usual pattern, having been designed by Repton.
"Instead of raising the surface to the level of the street,
as had usually been the custom, by bringing earth from
a distance," he "recommended a valley to be formed
through its whole length, with other lesser valleys flowing from it, and hills to be raised by the ground so
taken from the valleys." The original intention was to
bring the overflow of the Serpentine down Repton's

valley, but this was never done, and the gardens now only show the variation of level in one part. There is a good assortment of trees, and a group of mulberries which bear fruit every year.

Further west again, the old hamlet of Brompton has small, quiet squares of its own. The trees of Brompton Square, that quiet *cul-de-sac*, and the way through with a nice row of trees to Holy Trinity Church (built in 1829), with Cottage Place running parallel with it, is rather unlike any other corner of London. Before it was built over Brompton was famous for its gardens—first that of London and Wise, in the reign of William III. and Anne, and then that of William Curtis, the editor of the *Botanical Magazine*. A guide-book of 1792, describes Brompton as " a populous hamlet of Kensington, adjoining Knightsbridge, remarkable for the salubrity of its air. This place was the residence of Oliver Cromwell." Kensington Square is older than any of the Brompton Squares, having been begun in James II.'s reign, and completed after William III. was living in Kensington Palace. From the first it was very fashionable, and has many celebrated names connected with it—Addison, Talleyrand, Archbishop Herring, John Stuart Mill, and many others. The weeping ash trees and circular beds give the gardens a character of their own. Edwardes differs from all other London Squares. The small houses and large square garden are said by Leigh Hunt, who lived there at one time, to have been laid out to suit the taste of French refugees, who it was thought might take up their quarters there. The small houses were to suit their empty pockets, and the large garden their taste for a sociable out-of-door life. Loudon was an admirer of the design of the garden,

which he says was made by Aiglio, an eminent landscape painter, in 1819. The arrangement is quite distinct
from other squares—small paths, partly hidden by
groups of bushes and larger trees, all round the edge,
and from them twisting walks diverge towards the
centre. At their meeting-point now stands a shell
from the battle of Alma. The Square with its nice
trees, standard hollies, and even a few conifers and
carefully-planted beds, is further original in possessing
a beadle. This gentleman, who lives in a delightful
little house, with a portico in which the visitors to
the Square can shelter from the rain, looks most im-
posing in his uniform and gold-braided hat, and adds
greatly to the old-world appearance of the place. It
is sad to think the leases all fall in within the next
few years, and this quaint personage and vast garden
(it is 3¼ acres) and funny little houses may all disappear from London.

It is impossible in such a hasty glance to give
more than a very faint sketch of the story of the
squares, or a mere suggestion of the romance attached
to them. Though the gardening in many leaves much
to be desired, it is well to appreciate things as they
are, and enjoy to the full the pleasure the sight of
the huge planes in Berkeley or Bedford Squares, or
Lincoln's Inn Fields, can bring even to the harassed
Londoner. When the sun shines through the large
leaves, and the chequered light and shade play on the
grass beneath, and sunbeams even light up the massive
black stems, which defy the injurious fogs, they possess
a soothing and refreshing power. They, indeed, add to
the enjoyment, the health, and the beauty of London.

Q

CHAPTER X
BURIAL-GROUNDS

Praises on tombs are trifles vainly spent,
A man's good name is his best monument.
—Epitaph in St. Botolph, Aldersgate.

HE disused burial-grounds within the London area must now be counted among its gardens. There are those who would not have the living benefit by these hallowed spots set apart for the dead, but the vast majority of people have welcomed the movement which has led to this change. In some instances there is no doubt the transformation has been badly done. Here and there graves have been disturbed and tombstones heedlessly moved, but on the whole the improvement of the last fifty years has been immense. It is appalling even to read the accounts of many of the London grave-yards before this reaction set in. The hideous sights, the foul condition in which God's acre was often allowed to remain, as revealed by the inquiry held about 1850, together with the horrors of body-snatchers, are such a disagreeable contrast to the orderly graveyards of to-day, that the removal of a few head-stones is a much lesser evil.

Loudon, in the *Botanical Magazine*, was one of

the first to write about the improvement of public cemeteries, and to point out how they could be beautified, and the suggestion that the smaller burial-grounds could be turned into gardens was made as early as 1843 by Sir Edwin Chadwick. But the closing of them did not come until ten years later, and it was many years after that, before any attempt was made to turn them into gardens. By 1877 eight had been transformed, and from that time onwards, every year something has been done. The Metropolitan Gardens Association, started by Lord Meath (then Lord Brabazon) in 1882, has done much towards accomplishing this work. One of the earliest churchyards taken in hand was that of St. Pancras, and joined to it St. Giles-in-the-Fields. The Act permitting this was in 1875. Perhaps because it was one of the first, it is also one of the worst in taste and arrangement. The church of St. Pancras-in-the-Fields is one of the oldest in Middlesex. "For the antiquity thereof" it "is thought not to yield to St. Paul's in London." In 1593 the houses standing near this old Norman church were much "decaied, leaving poore Pancras without companie or comfort." The bell of St. Pancras Church was said to be the last tolled in England at the time of the Reformation, to call people to Mass. In the seventeenth and eighteenth century, adjoining to the south side of the churchyard, was "a good spaw, whose water is of a sweet taste," very clear, and imbued with various medicinal qualities. These "Pancras Wells" had a large garden, which extended from the Spa buildings by the churchyard, between the coach road from Hampstead, and the footpath across the meadows to Gray's Inn. As late as 1772 the coach was stopped and robbed at this corner, and

the footpads, armed with cutlasses, made off through
the churchyard. It was of this then lonely, rural churchyard that it was said the dead would rest "as secure
against the day of resurrection as . . . in stately Paules";
but, alas for modern exigencies, the Midland Railway
now spans the sacred ground by a viaduct, and the
would-be improvers, in turning what remained into a
garden, have moved the tombstones, levelled the undulating ground, and heaped the head-stones into
terrible rocky mounds, or pushed them in rows along
the wall. Numerous were the interesting monuments it
contained; many a courtly French *emigré* here found
a resting-place, such as the Comte de Front, on whose
tomb was the line, "A foreign land preserves his ashes
with respect." Although a monumental tablet put up
to record the opening, and the names of the designers
of the garden, proclaims it to be "a boon to the living,
a grace to the dead"; it is doubtful how that respect
to the dead was shown. The lines go on to say it was
"not for the culture of health only, but also of
thought." Surely health and thought could have been
equally well stimulated by making pretty paths, lined
with trees and flowers, wind reverently in and out
among the tombs, and up and down the undulating
ground, with seats in shade or sun, arranged with peeps
of the old church; and there might even have been
room for the fine sun-dial (the gift of Baroness
Burdett-Coutts) without levelling the whole area and
laying it out with geometrically straight asphalt walks.
The asphalt paths are in themselves a necessity in
most cases, as the expense of keeping gravel in order
is too great, and the majority of the renovated disused burial-grounds suffer from this fact.

Westward from St. Pancras the next large church-yard is that of Marylebone, and further to the north is St. John's Wood burial-ground. Its large trees and shaded walks are familiar to the thousands who go every year to Lord's Cricket Ground. Another large one, still more westward, now used as a garden, is Paddington. The small green patch round St. Mary's Church, and a large cemetery beyond, together make over 4 acres. All round London these spaces are being used, and in most cases little has been done to upset the ground—among the more prominent are St. George's, Hanover Square, in Bayswater; St. John's, Waterloo Road; Brixton Parish Church, with a row of yew trees; Fulham Parish Church, with Irish yews, and tall, closely clipped hollies; St. Mary's, Upper Street, Islington, and many others. Some are large spaces, such as St. John-at-Hackney, which covers 3 acres, and in it stands the tower of the old church, the present very large church which dominates it being in the Georgian style of 1797.

Stepney is the largest of all these disused church-yards, and covers 7 acres. It was opened as a public garden in 1887. The beautiful old Perpendicular church of St. Dunstan, with its carved gargoyles and fine old tower, which escaped the fire that destroyed the roof, stands on a low level, with the large square stone graves, of which there are a great quantity, on higher mounds round it. The central path, the old approach to the church, has trees on either side, and runs straight across the graveyard, and is as peaceful-looking as the walk in many a country churchyard. The way the lay-ing out as a garden has been carried out is unfortunate in many respects. The number of the big, stone, box-like monuments made it difficult to carry intersecting paths across between them, so a plan hardly to be commended has been followed, of half burying a number of these, and planting bushes in the earth thus thrown about, and putting the necessary frames for raising plants in the centre. To place the frames against the wall, and make a raised path or terrace among the tombs, and not to have banked them up with a kind of rockery of broken pieces, might have been more fitting. The part of the ground which is less crowded is well planted. Birch and alder (*Alnus cordifolia*) are doing well, and a nice clump of gorse flourishes.

One of the best-arranged of these old East End graveyards is that of St. George's-in-the-East, near Ratcliffe Highway. It is kept up by the Borough of Stepney, having been put in order under the direction of the rector, Rev. C. H. Turner (now Bishop of Islington), at the expense of Mr. A. G. Crowder, in 1866. The tombstones have for the most part been placed against the wall, or left standing if out of the way, as in the case of the one to the Marr family, whose murder caused horror in 1811. In the centre stands the obelisk monument to Mrs. Raine, a benefactress of the parish, who died in 1725. The whole of the ground is laid out with great taste and simplicity, and is thoroughly well cared for. The flowers seem to flourish particularly well, and the borders in summer are redolent with the scent of old clove carnations, which are actually raised and kept from year to year on the premises. A small green-house supplies the needs of the flower-beds. The superintendence of the garden is left to Miss Kate Hall, who takes charge of the Borough of Stepney Museum in Whitechapel Road, and also of the charming little nature-study museum in the St. George's Churchyard Garden. What formerly was the mortuary has been turned to good account, and hundreds of children in the borough benefit by Miss Hall's instruction. Aquaria both for fresh-water fish and shells, and salt-water collections, with a lobster, starfish, sea anemones, and growing sea weeds are to be seen, and moths, butterflies, dragon-flies, pass through all their stages, while toads, frogs, and salamanders and such-like are a great delight. The hedgehog spends his summer in the garden, and hibernates comfortably in the museum. The bees at work in the glass hive are another source of instruction. Outside the museum a special plot is tended by the pupils, who are allowed in turn to work, dig, and prune, and who obtain, under the eye of their sympathetic teacher, most creditable results. The charm of this East End garden, and the special educational uses it has been put to, shows what can be achieved, and sets a good example to others, where similar opportunities exist. A less promising neighbourhood for gardening could hardly be imagined, which surely shows that no one need be disheartened.

Some of the burial-grounds were in such a shocking state before they were taken in hand, that very few of the head-stones remained in their right places, and many had gone altogether, while some even reappeared as paving-stones in the district. Spa Fields, Clerkenwell, had a very chequered history. The site was first a tea garden, near the famous Sadler's Wells. For a few years, from 1770, its "little Pantheon" and pretty garden, with a pond or "canal" stocked with fish, and alcoves for tea drinkers, was thronged by the middle class, small tradesmen, and apprentices, while the more fashionable world flocked to Ranelagh or Almack's. It was the sort of place in which John Gilpin and his spouse might have amused themselves, on a less important holiday than their wedding anniversary. Twenty years later the scene had changed. The rotunda was turned into a chapel, by the Countess of Huntingdon, who took up her residence in a jessamine-covered house that had been a tavern, near to it. The gardens had already been turned into a private burial-ground, which soon became notorious for the evil condition in which it was kept. There every single gravestone had disappeared long before it was converted into the neat little garden, the delight of poor Clerkenwell children. The rotunda was at length pulled down, and in 1888 a new church was erected on the site. The same disgraceful story of neglect and repulsive overcrowding, can be told of the Victoria Park Cemetery, although the ground had not such a strange early history. It was one of those private cemeteries which the legislation with regard to other burial-places did not touch. It was never consecrated, and abuses of every kind were connected with it. It is a space of 9½ acres in a crowded district between Bethnal Green and Bow, a little to the south of Victoria Park. After various difficulties in raising funds and so forth, it was laid out by the Metropolitan Gardens Association, opened to the public in 1894, and is kept up by the London County Council, and is an extremely popular recreation ground, under the name of " Meath Gardens."

One of the quiet spots near the City is Bunhill Fields. This has for over two hundred years been the Nonconformist burial-ground. The land was enclosed by a brick wall, by the City of London in 1665 for interments "in that dreadful year of Pestilence. However, it not being made use of on that occasion," a man called "Tindal took a lease thereof, and converted it into a burial-ground for the use of Dissenters." As late as 1756 it appears to have been known as "Tindal's Burial-ground." The name Bunhill Fields was given to that part of Finsbury Fields, on to which quantities of bones were taken from St. Paul's in 1549. It is said "above a thousand cart-loads of human bones" were deposited there. No wonder the ghastly name of "bone hill," corrupted into Bunhill, has clung to the place. At the present time the gravestones here are undisturbed, and more respect has been shown to them than to the bones in the six-teenth century. Asphalt paths meander through a forest of monuments, and a few seats are placed in the shade of some of the trees. Those who live in this poor and busy district no doubt make much use of these places of rest, but the visitor is only brought to this depressing, gloomy spot on a pilgrimage to the tomb of John Bunyan. He rests near the centre of the ground, under a modern effigy. Not far off is the tomb of Dr. Isaac Watts, whose hymns are repeated wherever the British tongue is spoken, and near him lies the author of "Robinson Crusoe," Daniel Defoe. This quaint old enclosure opens off the City Road, opposite Wesley's Chapel, and on the western side it is skirted by Bunhill Row. But a few yards distant is another graveyard of very different aspect, as it contains only one stone, and that a very small one, with the name of George Fox, who died in 1690. The other graves in this, the "Friends' Burial-ground," never having been marked in any way, it has the appearance of a dismal little garden, like the approach or "gravel sweep" to a suburban villa. But it is neatly kept.

Of all the churchyards, that of St. Paul's is best
known, and least like the ordinary idea of one. But
this was not always so. It was for centuries an actual
burying-place. When the foundations of the present
cathedral were dug, after the Great Fire, a series of early
burials were disclosed. There were Saxon coffins, and
below them British graves, where wooden and ivory pins
were found, which fastened the woollen shrouds of those
who rested there, and below that again, between twenty
and thirty feet deep, were Roman remains, with fragments of pottery, rings, beads, and such-like.

The original churchyard was very much larger, as the
present houses in "St. Paul's Churchyard" are actually
on part of the ground included in it. It extended from
Old Change in Cheapside to Paternoster Row, and on
the south to Carter Lane, and the whole was surrounded
by a wall built in 1109, with the principal gateway opening into "Ludgate Street." This wall seems to have
been unfinished, or else part of it became ruinous in
course of time, and the churchyard became the resort
of thieves and ruffians. To remedy this state of things,
the wall was completed and fortified early in the fourteenth century. It had six gates, and remained like this
until the Great Fire, although long before that date
houses had been built against the wall both within and
without. Round here were collected the shops of the
most famous booksellers, such as John Day, who came
here in 1575.

On the north side was a plot of ground known as
Pardon's Churchyard, and here was built a cloister in
Henry V.'s time, decorated with paintings to illus-
trate Lidgate's translation of "The Dance of Death."
Here, too, was a chapel and charnel-house, and the

ST. PAUL'S CHURCHYARD

whole was pulled down by order of the Protector
Somerset, who used some of the material in building
Somerset House. It was on that occasion that the
cartloads of bones were removed to Finsbury Fields.
There, covered with earth, they made a solid, conspicuous hill on which windmills were erected. It was
part of this same ground which has already been referred
to as Bunhill Fields. Great as was the damage done by
the Fire, perhaps no site has been so completely altered as
that of St. Paul's. The modern cathedral, dearly loved
by all Londoners, stands at quite a different angle from
the old one, the western limit of which is marked by the
statue of Queen Anne. Nestling close to the south-west
corner of the great Gothic cathedral with its lofty spire,
was the parish church of St. Gregory, and the crypt was
the parish church of St. Faith's. Both these parishes
were allocated a portion of the churchyard for their
burials.

To the north-east of the cathedral stood Paul's
Cross, the out-door pulpit whence many notable sermons
were preached. It is described by Stowe. "About the
middest of this Churchyard is a pulpit-crosse of timber,
mounted upon steps of stone, and covered with Lead, in
which are Sermons preached by learned Divines, every
Sunday in the fore-noone. The very antiquity of which
Crosse is to me unknowne." The earliest scene he
records as taking place at this "crosse," was when
Henry III., in 1259, commanded the Mayor to cause
"every stripling of twelve years of age and upward to
assemble there," to swear "to be true to the King and
his heires, Kings of England." In later times, the most
distinguished preachers of the day were summoned to
preach before the Court and the Mayor, Aldermen and

citizens, and the political significance of such harangues may well be imagined. It was here Papal Bulls were promulgated; here Tyndal's translation of the New Testament was publicly burnt; here Queen Elizabeth listened to a sermon of thanksgiving on the defeat of the Armada—only to mention a few of the associations that cling round the spot, which, until within the last fifty years, was marked by an old elm tree which kept its memory green. Now it is treated with scant respect. There is, indeed, a little wooden notice-board, like a giant flower-label, stuck into the ground by an iron support, which records the fact that here stood Paul's Cross, destroyed by the Fire of 1666. The notice is not so large or conspicuous as the one a few feet from it, beseeching the kindly friends of the pigeons not to feed them on the flower-beds! It is to be hoped that before long the bequest of £5000 of the late H. C. Richards, for the re-erection of the Cross, may be embodied in some visible form.

What a picture such recollections call up!—the excited crowds with all the colour of Tudor costumes, the eager, fanatical faces of the "defenders of the Faith," the sad and despondent faces of the intensely serious Reformers, as they see the blue smoke curl upwards, and the flames consume the sacred volumes.

Picture the churchyard once more in still earlier times, when strange, fantastic customs clung round the cathedral services. One of the most original seems to have arisen from the tenure of land in Essex granted to Sir William Baud by the Dean and Chapter. The twenty-two acres of land were held on the condition that "hee would (for ever) upon the Feast day of the Conversion of Paul in Winter give unto them a good

Doe, seasonable and sweete, and upon the Feast of the Commemoration of St. Paul in Summer, a good Buck, and offer the same at the high Altar, the same to bee spent amongst the Canons residents." On the appointed days the keeper who had brought the deer carried it through the procession to the high altar. There the head was severed, and the body sent off to be cooked, while the horns, stuck on a spear, were carried round the cathedral. The procession consisted of the Dean and Chapter in their copes—special ones for the two occasions—one embroidered with does, the other with bucks, the gift of the Baud family, and on their heads garlands of roses. Having performed the ceremony within the church, the whole procession issued out of the west door, and there the keeper blew a blast upon his horn, and when he had "blowed the death of the Bucke," the "Horners that were about the City presently answered him in like manner." The Dean and Chapter paid the blowers of horns fourpence each and their dinner, while the man who brought the venison got five shillings and his food and lodgings, and a "loafe of bread, having the picture of Saint Paul upon it," to take away with him. What a strange picture of mediæval life and half-pagan rites! yet all conducted with perfect good faith, in all seriousness. It is just one of the great charms of knowing London and its traditions, that one is able to clear away in imagination the growth of centuries, and throw back one's mind to the past—to stand at the top of Ludgate Hill and to remove Wren's building and to see the Gothic pinnacles; to blot out the garden and fountain and modern seats, and see Paul's Cross; on the left to see the arches of the cloisters, and on the right the high wall and timbered houses; then to open the western door and see this strange procession issue forth, with the antlers borne aloft, and hear the bugle-blast and answering notes.

Surely no place can be more crowded with memories than busy, "roaring London," and nowhere are the past and present so unexpectedly brought together. The City is full of surprises to those who have leisure to wander among its narrow, crowded streets. The quiet little graveyards afford many of these telling contrasts. Suddenly, in the busiest thoroughfares, where a constant stream of men are walking by every week-day, come these quiet little back-waters. In many cases the churches themselves have vanished, or only remain in part. St. Mary's Staining is one of these, so hidden away that one might walk along Fenchurch Street hundreds of times and never find it. The approach is by a very narrow alley, at the end of which is this quiet little graveyard, where, among other worthies, reposes Sir Arthur Savage, knighted at Cadiz in 1596. The church, all except the tower, was destroyed in the Great Fire, and never rebuilt. The picturesque old tower stands in the centre of this little plot, which now forms the garden of the Clothworkers' Company, whose hall opens on to one side of it.

Another church which perished in the Fire and was never rebuilt is St. Olave's, Hart Street, but its church-yard remains, and a few large tombs stand in a small garden with seats, where at all times of the year some weary wayfarers are resting.

Another such graveyard where the burnt church was not restored is at the corner of Wood Street and Cheapside. The old tree inside the closed railings

may have inspired the lark to carol so joyously as to
call up the "vision of poor Susan."

St. Botolph's, Aldersgate, has one of the largest
churchyards in the City, but it really consists of four
pieces of land thrown into one in 1892, by a scheme
under the London Parochial Charities, which contributed part of the purchase-money of some of the
land, and gives £150 a year for the upkeep—£100
being paid to them by the General Post Office, which
has the right of light over the whole space. One-
half of the churchyard is St. Botolph's, and the rest
is made up of the burial-grounds of St. Leonard,
Foster Lane, and Christ Church, Newgate Street,
and a strip of land which might have been built
on, but which, under the revised scheme in 1900,
became permanently part of this open space. The
garden is carefully laid out ; there are nice plane trees
and a little fountain, regular paths and numerous seats.
A sheltered gallery runs along one side, and in it
are tablets to commemorate deeds of heroism in humble
life—Londoners who lost their lives in saving the lives
of others. The church of St. Botolph was one which
escaped the Fire, but had fallen into such disrepair
that it was rebuilt, by Act of Parliament, in 1754.
The Act specially stipulates that none of the gravestones were to be removed, but where some of them
are, now that it is a trim garden, it would be hard to
say. Being not far from the General Post Office, this
garden is so much used by its officials during the middle
of the day, it has earned the name of the " Postman's
Park."

Another much-frequented but much smaller churchyard is that of St. Katharine Coleman. Suddenly, in

SUN-DIAL, ST. BOTOLPH'S

a corner of crowded Fenchurch Street, comes this retired shade. The church, with its old high pews, and tiny graveyard, devoid of monuments, is a peaceful oasis. These surprises in the densest parts of the City are very refreshing, and they are too numerous to mention each individually. Most of them now are neatly kept, though some look dreary enough. None of them recall the neglect of half a century ago. St. Olave's, Hart Street, in Seething Lane, is perhaps among the most gloomy. It is the church Pepys speaks of so often as "our owne church," and was one of the churches that escaped the Fire. The archway with the skulls over it, leads from Seething Lane to the dismal-looking churchyard. Nothing is done to alter or brighten this place of many memories. One shudders to think of what it must have been like when Pepys crossed it for the first time after the Great Plague, when he went to the memorial service for King Charles I., on 30th January 1666. No wonder he says it "frighted me indeed to go through the church more than I thought it could have done, to see so many graves lie so high upon the churchyard, where people have been buried of the Plague. I was much troubled about it, and do not think to go through it again a good while." The parish registers show that no less than 326 were interred in this very small place, during the previous six months, so Pepys' feelings were well justified. The old church has a special interest to lovers of gardens, as in it is the tomb of William Turner, the author of the first English Herbal.

In more than one City churchyard a portion of the old wall makes its appearance. There is St.

R

Alphage, London Wall, and Allhallows-in-the-Wall,
where the little gardens by the wall have been formed
with a view to preserving it. The most picturesque
is St. Giles's, Cripplegate, where Milton is buried. The
graveyard is large, and the ground rises above the
footpath, which was made across it some thirty years
ago, to a bastion of the wall, of rough stones and
flint, which is in its old state, although part of the
wall was rebuilt in 1803. There has been no attempt
here to make it a resting-place for the living, although
it is used as a thoroughfare.

Few people who have not entered the Bank of England would suspect it of enclosing an extremely pretty
garden. There the inner courtyard possesses tall lime
trees, gay rhododendrons, and a cool splashing fountain,
with ferns and iris glistening in the spray. It is quite
one of the most delightfully fresh and peaceful corners
on a hot summer's day, and carries one in imagination to
Italy. Yet this is but another of the many old City
churchyards. The parish of St. Christopher-le-Stocks
was absorbed, with five other parishes, into St. Margaret's,
Lothbury, in 1781. Some of the tombs, and pictures of
Moses and Aaron, were removed from it, and are still
to be seen in St. Margaret's, which is crowded with
monuments from all six churches. The Bank was
already in possession of most of the land within the
parish, and by the Act of Parliament of 1781, the church
and churchyard became part of the Bank premises, which
cover nearly three acres. The church site was built
over, but the graveyard became the garden. This
enclosure at first was a simple grass plot, as shown in
an engraving dated 1790. The lime trees may have
been planted soon after, as they appear as large trees

THE BANK GARDEN

sixty years later, and are spoken of in 1855 as two of
the finest lime trees in London. The fountain was put
up in 1852 by Mr. Thomas Hankey, then the governor.
The water for it came from the tanks belonging to the
Bank, supplied by an artesian well 330 feet deep, said to
be very pure, and free from lime. Perhaps that is why
the rhododendrons look so flourishing. Most of the
Bank, as is well known, was the work of the architect Sir
John Soane, but some of the portions built by Sir Robert
Taylor, before his death in 1788, when Soane was appointed to succeed him, are to be seen in the garden
court. It is said that the last person buried there
was a Bank clerk named Jenkins, who was 7½ feet
in height. He was allowed to rest there, as he feared
he might be disinterred on account of his gigantic
proportions.

Very different is the churchyard of St. Martin's, on
Ludgate Hill. It belongs to Stationers' Hall, and
although it boasts of one fine plane tree, is an untidy,
grimy, dingy little square. By permission of all the
necessary authorities, the coffins (480 in number) were
removed and reverently buried in Brookwood Cemetery
in 1893, a careful register of all the names and dates,
that could be deciphered, being kept. This having
been done, the earth was merely left in an irregular
heap round the tree, and no attempt has been made
to improve in any way the forsaken appearance of the
place.

This sketch does not aim at being a guide-book,
and it would only be tedious to enumerate the many
churchyards, without as well as within the City, which
of late years have been made worthy " gardens of sleep."
St. Luke's, Old Street; St. Leonard's, Shoreditch;

St. Anne's, Soho; St. Sepulchre, Holborn, and many others in every part of the town, from being dreary and untidy, have become orderly and well kept; and instead of being unwholesome and unsightly, have become attractive harbours of refuge in the sea of streets and houses.

CHAPTER XI
INNS OF COURT

Sweete Themmes ! runne softly, till I end my Song.
At length they all to mery London came,

. . . .

There when they came, whereas those brickly towers
The which on Themmes brode aged backe doe ryde
Where now the studious Lawyers have their bowers,
There whylome wont the Templer Knights to byde,
Till they decayed through pride :

. . . .

Sweete Themmes ! runne softly, till I end my Song.
—Spenser : "Prothalamion, or a Spousall Verse."

HERE are no more peaceful gardens in all London than those among the venerable buildings devoted to the study of the law. There is a sense of dignity and repose, the moment one has entered from the noisy thoroughfares which surround these quiet courts. They may be dark, dull, and dingy, as seen by a Dickens, and sombre and serious, to those whose business lies there; but to the ordinary Londoner, who loves the old world of the City, and the links that bind the present with the past, there are no more reposeful places than these gardens. The courts and buildings seem peopled with

those who have worked and lived there. If stones could speak, what tales some of these could tell!

The best-known, perhaps, of the gardens are those belonging to the Inner and Middle Temple, as their green lawns are visible from the Embankment. They add greatly to the charm of one of London's most beautiful roadways, now, alas! desecrated by the rush of electric trams, and its fine young trees sacrificed to make yet more rapid the stream of beings hourly passing between South London and the City. The modern whirl of business life can leave nothing untouched in this age of bustle, money-making, ceaseless toil, and care. Even pleasures have to be provided by united effort, and partake of noise and hurry. Thought and contemplation are hardly counted among the pleasures of life; yet to those who value them, even to look through the iron railings on the smooth turf brings a sense of relief. Even to those who scarcely seem to feel it, the very existence of these haunts of comparative peace, which flash on their vision as they hurry by, leaves something, a subtle influence, a faint impression on the brain. It must make a difference to a child who knows nothing beyond the noisy streets and alleys in which its lot is cast, to hear the rooks caw and the birds sing in the quiet gardens of Gray's Inn. It must come as a welcome relief, even though unperceived and unappreciated, from the din and clatter in which most of its days are passed. One cannot be too grateful that it has not been thought necessary to change and modernise "our English juridical university."

Although the four great Inns of Court are untouched, the lesser Inns have vanished or are vanishing. Clement's Inn has gone. The garden there was small, but had a special feature of its own—a sun-dial upheld by the kneeling figure of a blackamoor. This is now preserved in the Temple Garden, where it appeared soon after Clement's Inn was disestablished in 1884. Clement's Inn, which appertained to the Inner Temple, was so named from the Church of St. Clement Danes and St. Clement's Well, where "the City Youth on Festival Days used to entertain themselves with a variety of Diversions." The sundial is said to have been presented to the Inn by a Holles, Lord Clare, and some writers state that it was brought from Italy. It was, however, more probably made in London by John Van Nost, a Dutch sculptor, who came to England in William III.'s time, and established himself in Piccadilly. When he died in 1711 the business was continued by John Cheere, brother of Sir Henry Cheere, who executed various monuments in Westminster Abbey. Similar work is known to have issued from this studio. At Clifford's Inn, which was also attached to the Inner Temple, there is still a vestige of the garden, but it looks a miserable doomed wreck, a few black trees rising among heaps of earth and rubbish. It was described in 1756 as "an airy place, and neatly kept; the garden being inclosed with a pallisado Pale, and adorned with Rows of Lime trees, set round the gravel Plats and gravel walks." Its present forlorn appearance is certainly not suggestive of its past glories. Barnard's Inn has been converted into a school by the Mercers' Company; it also has its court and trees on a very small scale. Staples Inn, so familiar from the timbered, gabled front it presents to Holborn, carefully preserved by the Prudential Assurance Company, its present owners, still has its quiet little quadrangle of green at the back. It was of that Dickens wrote such an inimitable description. "It is one of those nooks, the turning into which out of the clashing streets imparts to the relieved pedestrian the sensation of having put cotton in his ears and velvet soles on his boots." Furnival's, Thavies', and all the other Inns famous in olden days, have disappeared, and their quiet little gardens with them.

The Temple Gardens are larger now than in the earlier days of their history, as then there was nothing to keep the Thames within its channel at high tide. The landing steps from the river were approached by a causeway of arches across the muddy banks. It was not until 1528 that a protecting wall was built, and a pathway ran outside the wall between it and the river. Gardens must have existed on this site from a very early date. When the Templars moved there from Holborn and built the church in 1185, it was all open country round, with a few great houses and conventual buildings standing in their own orchards and gardens. After the suppression of the Order, it was in the hands of Aimer de Valence, Earl of Pembroke, and in 1324 the land was given to the Knights of St. John. As they had their own buildings and church not far off, they granted it " to the Students of the Common Lawes of England : in whose possession the same hath sithence remained." All the consecrated land, and all within the City, was included in the grant to the Knights of St. John : besides this there was some land outside the City, or the Outer Temple, part of which remained in secular hands, and in later times was covered by Essex House, with its famous gardens. The section belonging to the Law Societies, beyond the City, is spoken of in early records as the Outer Garden, and from time to time buildings were erected on it—at first under protest,

as in 1565 there was an order " for the plucking down
of a study newly erected," and again in 1567, " the
nuisance made by Woodye, by building his house in
the Outer Garden, shall be abated and plucked down,
or as much thereof as is upon Temple ground." All
this garden has long ago been completely built over,
and the large spaces now forming the Temple Gardens
are those anciently known as the " Great Garden,"
belonging to the Inner Temple and the Middle Temple
Garden. The Outer Temple (never another Inn) was
merely the ground outside the limits of the City.

The long green slopes down to the Embankment,
are much larger than the older gardens, as the wall
which was built in 1528 to keep out the river, cut
across from where No. 10 King's Bench Walk now
stands. The wall must have been a vast improvement,
and was greatly appreciated. In 1534 a vote of thanks
was passed by the " parliament " of the Inner Temple
to the late Treasurer, John Parkynton, who had " takyn
many and sundrie payns in the buylding of the walle
betwene the Thamez and the garden," for which
" greate dyligens " they gave unto him " hartey thankes."
And, indeed, the garden must sorely have needed this
protection. It is difficult to picture the Temple in
the sixteenth century, and the little gardens must
have been as bewildering as the present courts and
buildings. In the records there are references to various
gardens, no doubt small enclosures like the present
courts, besides the Great Garden and the kitchen-
garden. There was the nut garden, perhaps adorned
with nut trees, as Fig-tree Court probably was with
figs. There is more than one record of payments for
attending to the fig-tree or painting rails round it.

In 1610, just at the time James I. brought them into
notice, a mulberry was "set in Fairfield's Court." In
1605 seats were set "about the trees in Hare's Court";
thus all the courts were more or less little gardens.
In 1510 a chamber is assigned to some one "in the
garden called le Olyvaunte." This was probably the
Elephant, from a sign carved or painted to distinguish
a particular house facing it. There was similarly "le
Talbott," probably from a greyhound sign, in another
court. The houses facing the Great Garden apparently
had steps descending into it from the chief rooms,
and it was a special privilege to have your staircase
opening on to it. Thus, "May 1573, Mr. Wyott and
Mr. Hall, licensed to have 'a steeyrs' (stairs) from
their chamber into the garden." The Great Garden
was constantly being encroached on as new chambers
were built. Entries in the records with regard to permission to build into the garden often occur; for
instance—

"1581. Thomas Compton . . . to build . . . within
the compass of the garden or little Court . . . from the
south corner of the brick wall of the said garden . . .
57 feet . . . and from the said wall into the garden
22 feet."

On one occasion a license to build was exceeded, and
the offence further aggravated by cutting down "divers
timber trees." The offender was at first put out of
commons, and fined £20, which was afterwards mitigated
to £5, with the addition of a most wise proviso, that
"he shall plant double the number of trees he caused
to be cut down." Would that the fault of felling
timber always met with the same punishment!

When houses were put on the site of the present

Paper Buildings in 1610, the Great Garden was cut in two, and the eastern portion went to form the broad stretch with its trees known as King's Bench Walk. Elm trees were planted, and the walks and seats under them repaired from time to time, and kept in good order. The part to the west was carefully tended, and became from that year the chief garden. In James I.'s reign, that age of gardening, when every house of any pretensions was having its garden enlarged, and Bacon was laying out the grounds of Gray's Inn, the Temple was not behind-hand. The accounts show constant repairs and additions and buying of trees. The items for painting posts and rails are very frequent. Probably they do not always refer to outer palings, but it may be that the Tudor fashion of railing round the beds, with a low trellis and posts at the angles, still prevailed. One of the largest items of the expenses was for making "the pound" in 1618. This, it is said, was a pond, but no record of digging it out, or filling it with water occurs, while all the payments in connection with it went to painters or carpenters, and therefore it was more probably a kind of garden-house, much in favour at that time, made by the wall, to command a view over the river. The chief items with regard to it are:—

"1618. To John Fielde, the carpenter, for making ' the pound ' in the garden, £19."

" To Bowden, the painter, for stopping and 'refreshing' the rails in the ' wakes ' (walks), the posts, seats and balusters belonging to the same, and for stopping and finishing the ' pound ' by the waterside, £9, 10s."

Again in 1639 the entry certainly implies some kind of summer-house and not "a pond " : " Edward Simmes, carpenter, for repairing 'the pound' and other seats in the garden and walks, &c., £15, 8s." There must have been another summer-house at the same time, unless the sums paid to a plasterer "for work done about the summer-house in the garden," in 1630, refers to the same "pound."

A great deal seems to have been done to the Garden during the first few years of the Commonwealth, and large sums were expended in procuring new gravel and turf: "392 loads of gravel at 2s. 6d. the load" is one entry. But the chief work was the re-turfing. An arrangement was made, by payment of various small sums to the poor of Greenwich, to cut 3000 turfs on Blackheath, and convey them in lighters to the Temple Stairs. A second transaction procured them 2000 more, each turf being a foot broad and a yard long. These amounts would cover a third of an acre with turf. The head gardeners seem to have been particularly unruly people. Although they remained in office many years, there were frequent complaints. On one occasion this official had cut down trees, another time he had the plague, and his house was frequented by rogues and beggars. At first the gardener's house was on the present King's Bench Walk side of the Garden, near the river; later on, near where Harcourt Buildings are now. In 1690 the house, then in Middle Temple Lane, was turned into an ale-house, and evidently none of the quietest, for the occupier was forbidden to sell drink, and the "door out of the gardener's lodge towards the water gate" was ordered to be bricked up, so as to prevent all the riffraff from the river rioting in his rooms. Yet the post descended from father to son. In 1687 Thomas Elliott succeeded his father,

Seth Elliott, who had been there some years, and when
in 1708 Charles Gardner had taken the second Elliott's
place, his daughter Elizabeth's name occurs as a recipient
of money, and Elliott himself received a pension of £20
a-year, although he was the culprit of the riotous ale-
house. During the years succeeding the Restoration, the
Garden seems to have been little touched. The kitchen-
garden would still be maintained, and either it was
farmed by the gardener, or its supplies were inadequate,
as on fast-days there was always a special payment to the
gardener for vegetables. Such items as the following are
of frequent occurrence : " Sallating for the hall in grass
week, strewings and ' bow pots' for the hall in Easter
and Trinity terms."

Though the French fashions in gardening of Charles
II.'s reign do not seem to have affected the Temple precincts, yet the Dutch influence that came in with
William and Mary made itself felt. A small garden
was specially set apart for the Benchers, and done up
entirely in the prevailing style. A piece of ground
between King's Bench Office and Serjeants' Inn was made
use of for this. It had been let to the Alienation Office,
but after the Great Fire the Temple resumed the control
of it, and finally did it up and replanted it for the use
of the Benchers. It was known as the " Benchers'," the
" Little " or the " Privy " Garden, and great care,
attention, and money were expended on it. Turf,
gravel, and plants were bought; a sun-dial put on
the wall ; orange trees set out in tubs ; and a fountain
erected in the middle. This fountain must have been
the chief feature of the Garden, and from the immense
amount of care it required to keep it in order, it seems
that it was one of those elaborate " waterworks," with-

out which no garden was then complete. Such fountains were made with secret arrangements for turning on the water, which dropped from birds' bills, or spurted out of dolphins or such-like, with an unpleasant suddenness which gave the unwary visitor a shower-bath. Other fountains played tunes or set curious machinery in motion, or otherwise surprised the beholder. From the descriptions, this one in the Benchers' Garden doubtless concealed some original variation. It consisted of a lion's face with a copper scallop shell, and a copper cherry-tree with branches, and perhaps the water dropped from the leaves. One payment in 1700 occurs for " a new scallop shell to the fountain, for a cock and a lion's face to draw the water out of the fountain, and for keeping the fountain in repair, £12." The copper cherry-tree was painted, and perhaps the Pegasus—the arms of the Inner Temple—figured in the strange medley, as the cost of painting the tree and " gilding the horse" are together paid to the man " Fowler," who had charge of the fountain. The " best way to bring the water" had to be carefully considered for these " waterworks " which Fowler was designing and carrying out, and it evidently was brought up to the pitch of perfection required of a fountain in those days. There was also a summer-house with a paved floor, and an alcove with seats. Altogether, even without the glories of the strange fountain, the little enclosed Dutch garden must have been an attractive place.

While the Benchers' Garden was being made, the Great Garden was not neglected. Its form was altered to suit the prevailing taste. This remodelling must have begun in the winter of 1703, as it was then resolved that " the trees in the Great Garden be cut down, and

THE INNER TEMPLE GARDEN

the Garden to be put in the same model as the gardener hath proposed." The delightful terrace, which is still one of the most beautiful features in the Garden, existed before these alterations began, but the sun-dial which still adorns it was added during these changes. The payment for it was made to Strong, who was contractor for St. Paul's under Wren: "To Edward Strong, for the pedestal for the dial in the Great Garden steps, &c., £25." The beautiful gates of wrought iron were put up in 1730. The design shows the arms of Gray's Inn, as well as the winged horse of the Inner Temple, in compliment to the other learned society, its close ally. In the same way the Pegasus occurs at Gray's Inn. It was probably along this terrace that some of the orange trees in pots were placed during the summer. The pots in which these oranges and other "greens" were grown seem to have been specially decorative. It was a serious offence when Allgood, a member of the Inn, broke some, and was obliged to "furnish other pots of like fashion and value," otherwise he would "be put out of commons." After this others were purchased, as the payment of £8 was made "for a large mould, carved in wood, for casting of earthen pots for the Garden"; and in other years further similar expenses occur, one in 1690 "to the potter for a large pot made for the Garden, painted in oil, £1, 5s." Some of the plants grown would stand the winter in the open, but after the oranges made their appearance a shelter had to be provided. Green-houses owed their origin to this necessity, and as they were only used in winter, and merely sheltered the large pots of "greens," these green-houses or orangeries were built like rooms, and used as summer-houses during warm months. All the larger gardens had their green-houses, but the smaller proprietors frequently sent their plants away to a nurseryman to be housed during the winter. Even the "greens" at Kensington Palace were kept by London and Wise, until the new orangery was built. The Temple orange trees were first sent to the house of Cadrow at Islington. In 1704 the green-house seems to have been made, and used as a garden-house in summer. Such items in the accounts as "a chimney-glass and sconces for the green-house" show that it was in the usual solid architectural style then in fashion. That the "panierman," an officer, one of whose duties was to summon members to meals by blowing a horn, was appointed to take charge of it as well as of the library, is a further proof that it bore the character of a room, and was more or less outside the gardener's department. The panierman also had the care of the elaborate fountain, after it had been supervised for some years by the maker. This green-house stood at the end of the terrace, which still runs parallel with Crown Office Row, and near the site of Harcourt Buildings, behind the gardener's house. This gardener's house was pulled down two or three years later to make way for Harcourt Buildings, which was joined to the summer-house. The first or ground floor opened on to the garden below the "paved walk" or terrace, on which level stood the summer-house.

The most fascinating feature of a garden ought to be its flowers, and of these also some particulars can be gleaned from the accounts. There is enough to show that the Temple Garden was quite up to date in its horticulture, and that it followed fashion as closely in its plants as in its design. It is not surprising to find Dutch bulbs, and especially tulips, being bought when such a lover of those flowers as Sir Thomas Hanmer was a member. He was one of those who devoted much time to the culture of that flower, when the tulip mania was at its height, and raised new varieties, which were known by his name, "the agate Hanmer." In 1703 the list of bulbs purchased is carefully noted. There were "200 'junquiles' at 6s. a hundred; for 200 tulips at 5s. a hundred; for 100 yellow Dutch crocus, for 50 Armathagalum." The spelling of "junquiles" is much more correct than our modern "jonquil," and all the old writers would have written it so. Parkinson, in 1629, describes them as "Narcissus juncifolius" or the "Junquilia or Rush Daffodill"; but "Ornithogalum" was too much for the Temple scribe. The "Ornithogalum" or "Starre of Bethlehem," and probably one of the rarer varieties, must be meant by "Armathagalum." The Arabian variety was then "nursed in gardens," but it should be "housed all the winter, that so it may bee defended from the frosts," wrote Parkinson, and sadly admitted that the two roots sent to him "out of Spain" had "prospered not" "for want of knowledge" of this "rule." There was also the "Starre flower of Æthiopia," which "was gathered by some Hollanders on the West side of the Cape of Good Hope"; and this is more likely to have been the variety bought for the Temple with the other Dutch bulbs. Among the other purchases were various shrubs, on which the topiary art was then commonly practised. There were "15 yew trees for the Great Garden in pots, . . . 4 box trees for the grass plots, . . . 12 striped 'fillerayes' "—this latter being variegated phillyreas (most likely *angustifolia*), which were largely used for cutting into quaint shapes. Another account is for "28 standard laurels, 4 'perimic' (laurels),

6 junipers, 4 hollies, and 2 perimic box trees." These
"perimetric" trees had already gone through the necessary clipping and training, to enable them to take their
place in the trim Dutch garden. Another year flowering
shrubs are got for the Benchers' Garden: "2 messerius
at 2s., and 2 lorrestines at 2s." The *Daphne mezereum*
had been a favourite in English gardens from the earliest
times, and the laurestinus (*Viburnum tinus*) came from
South Europe in the sixteenth century. Parkinson, the
most attractive of all the old gardening authors, has a
delightfully true description of the "Laurus Tinus," with
its "many small white sweete-smelling flowers thrusting
together, . . . the edges whereof have a shew of a wash
purple or light blush in them ; which for the most part
fall away without bearing any perfect ripe fruit in our
countrey : yet sometimes it hath small black berries, as if
they were good, but are not"! Fruit-trees were also to
be found—peaches, "nectrons," cherries, and plums,
besides figs and mulberries. That the walls were
covered with climbing roses and jessamine is certain,
from the oft-recurring cost of nailing them up. "Nails
and list for the jessamy wall," and the needful bits of
old felt required to fasten them up, was another time
supplied by "hatt parings for the jessamines."

Thus it is easy, bit by bit, out of the old accounts,
to piece together the Garden, until the mind's eye can see
back into the days of Queen Anne, and take an imaginary
walk through it on a fine spring evening. The Bencher
walks out of the large window of the "green-house" on
to the terrace, where the sun-dial points the hour: the
orange trees, glossy and fresh from their winter quarters,
stand in stiff array, in the large artistic pots. Down the
steps, a few stiff beds are bright with Dutch bulbs in

flower. The turf, well rolled (for a new stone roller has just been purchased), stretches down to the river between straight lines of quaintly cut box, yews, and hollies. He sees Surrey hills clear in the early evening light, and the barges sail by, and boats pass up and down the river. He may linger on one of the seats in the garden-house overlooking the river, or wander back under the stately elms of King's Bench Walk, to rest awhile in the Privy Garden, where the air is scented with mezereum, and cooled by the drops that fall from the metal leaves hanging over the basin of the fountain.

The Middle Temple, too, had its Benchers' Garden, and part of it survives to this day in the delightful Fountain Court. The Benchers' Garden was larger, covering the ground where Garden Court now stands, up to the wall of the famous gardens of Essex House. A garden covered the space where the library has been built, and the terrace and steps in front of the fountain reached right across to the Essex House wall. Below the beautiful old hall which Queen Elizabeth opened in person, and where Shakespeare's contemporaries witnessed " Twelfth Night," lay the rest of the Garden, with green lawns and shady trees down the water's edge. The fountain, once the glory of the Benchers' private garden, is still one of the most delightful in all London. Sir Christopher Hatton, whose garden of Ely Place—wrung by Queen Elizabeth from the unwilling Bishop—was not far off, was an admirer of the Middle Temple fountain. It was kept, he says, "in so good order as always to force its stream to a vast and almost incredible altitude. It is fenced with timber palisades, constituting a quadrangle, wherein grow several lofty trees, and without are walks extending on every side of the quadrangle, all paved with Purbeck, very pleasant and delightful." In an eighteenth-century picture, with groups of strollers and a lady passing the gay company in her sedan chair, the palings are superseded by fine iron railings enclosing the lofty jet, its marble basin, and shady trees. The pavement ended with the terrace wall overlooking the garden below, and the Thames covered at high tide what is now the lower part of the lawn. The Fountain Court has inspired many a thought which has found expression in prose and verse, but no picture is more vivid or well known than the figure of Ruth Pinch, in " Martin Chuzzlewit," waiting for her brother " with the best little laugh upon her face that ever played in opposition to the fountain," or the description at the end, of that crowning day to her happiness, when she walks there with John Westlock, and " Brilliantly the Temple Fountain splashed in the sun, and laughingly its liquid music played, and merrily the idle drops of water danced and danced, and peeping out in sport among the trees, plunged lightly down to hide themselves, as little Ruth and her companion came towards it." The fountain has suffered some modernising changes since Dickens wrote those lines ; but in spite of them there is still music in its sound, which calls up dreams of other ages and of brighter gardens as it tosses its spray into the murky air.

> " Away in the distance is heard the vast sound
> From the streets of the city that compass it round,
> Like the echo of mountains or ocean's deep call :
> Yet that fountain's low singing is heard over all."
> —Miss Landon.

Of all the incidents that are associated with particular places, none stands out more vividly than the scene told

Victoria Manners. 1905.

THE FOUNTAIN COURT, MIDDLE TEMPLE

by Shakespeare, of the first beginning to the Wars of the
Roses in the Temple Garden.

Richard Plantagenet, with the Earls of Somerset,
Suffolk, and Warwick, Vernon, and a lawyer, enter the
Temple Garden (" Henry VI." Pt. I. Act 2, sc. iv.).

Suffolk. Within the Temple Hall we were too loud;
The garden here is more convenient.
Plantagenet. Then say at once if I maintained the truth,
Or else was wrangling Somerset in the error?

.   .   .   .   .   .

The direct answer being evaded, Plantagenet continues—

Since you are tongue-tied and so loath to speak,
In dumb significants proclaim your thoughts;
Let him that is a true-born gentleman,
And stands upon the honour of his birth,
If he suppose that I have pleaded truth,
From off this brier pluck a white rose with me.
Somerset. Let him that is no coward nor no flatterer,
But dare maintain the party of the truth,
Pluck a red rose from off this thorn with me.

.   .   .   .   .   .

Warwick. I pluck this white rose with Plantagenet.
Suffolk. I pluck this red rose with young Somerset.

.   .   .   .   .   .

Vernon. I pluck this pale and maiden blossom here,
Giving my verdict on the white rose side.

.   .   .   .   .   .

Lawyer (to *Somerset*) . . . The argument you held was
wrong in you,
In sign whereof I pluck a white rose too.
Plan. Now, Somerset, where is your argument?
Som. Here, in my scabbard, meditating that
Shall dye your white rose in a bloody red.

.   .   .   .   .   .

Plan. Hath not thy rose a canker, Somerset?
Som. Hath not thy rose a thorn, Plantagenet?

Plan. Ay, sharp and piercing to maintain his truth ;
Whiles thy consuming canker eats his falsehood.

Som. Well, I'll find friends to wear my bleeding roses,
That shall maintain what I have said is true.

. . . . . .

Warwick. And here I prophesy this brawl to-day,
Grown to this faction in the Temple-garden,
Shall send between the red rose and the white
A thousand souls to death and deadly night.

With such a tradition the Temple Garden should
never be without its roses. They are one of those
friendly plants which will do their best to fight against
fog and smoke, and flower boldly for two or three years
in succession: so a supply of red and white, and the
delightful *Rosa mundi*, the "York and Lancaster,"
could without much difficulty be seen there every
summer. Certainly some of the finest roses in existence
have been in the Temple Gardens, as the Flower Shows,
which are looked forward to by all lovers of horticulture,
have for many years been permitted to take place in these
historic grounds. How astonished those adherents of
the red or white roses would have been to see the colours,
shades, and forms which the descendants of those briars
now produce. The Plantagenet Garden would not contain many varieties, although every known one was
cherished in every garden, as roses have always been first
favourites. Besides the briars, dog roses, and sweet
briars, there was the double white and double red, a
variety of *Rosa gallica*. Many so-called old-fashioned
roses, such as the common monthly roses, came to England very much later, and the vast number of gorgeous
hybrids are absolutely new. Elizabethan gardens had a
fair show of roses with centifolia, including moss and
Provence roses, and York and Lancaster, *Rosa lutea*, musk,

damask, and cinnamon roses in several varieties; and as
the old records show, the Temple Garden was well supplied
with roses. All these probably flourished there in the
days of Shakespeare, and would readily suggest the scene
he immortalised.

Among the spirits that haunt the Temple Garden,
there is none that seems to cling to it more than that of
Charles Lamb. It should be a pride of these peaceful
gardens that they helped to mould that lovable and
unselfish character. A schoolfellow, who describes his
ways as a boy at Christ's Hospital, recalls how all his
young days were spent in the solemn surrounding of the
Temple, and how, while at school, "On every half holiday
(and there were two in the week), in ten minutes he was
in the gardens, on the terrace, or at the fountain of the
Temple. Here was his home, here his recreation; and
the influence they had on his infant mind is vividly
shown in his description of the old Benchers."

"Shadows we are and like shadows depart," suggests
the sun-dial on the wall of Pump Court, but shadows of
such gentle spirits as Charles Lamb leave something
behind, and those "footprints on the sands of time"
are nowhere more traceable than in these solemn
precincts of law with their quiet, restful gardens.

The attractions of the Temple are so great, one feels
loth to cross the noisy thoroughfare and plunge
through the traffic till the stately old gateway out of
Chancery Lane, on which Ben Jonson is said to have
worked, affords an opening towards the spacious gardens
of Lincoln's Inn.

Lincoln's Inn Gardens have a special claim to
antiquity as they are partly on the site of the famous
garden of the Earl of Lincoln, of which some of the

accounts are preserved in a splendid big old manor roll
now at the Record Office. It is supposed that at his
death in 1311, Henry de Lacy, Earl of Lincoln,
assigned these lands to the " Professors of the Law as a
residence." Additions were made later from the ground
belonging to the Bishop of Chichester, round the palace
which Ralph Neville had built in 1228. Part of the
site was the " coney garth," which belonged to one
William Cotterell, and hence is often mentioned as
" Cotterell's Garden." Garden of course only meant a
garth or yard, and though the name now signifies an
enclosure for plants, in early times other enclosures
were common. There was the "grass yard " or lawn,
the " cook's garth" or kitchen-garden, and " coney
garth " where rabbits were kept, as well as the " wyrt
yard " or plant yard, the " ort yard " or orchard, apple
yard, cherry yard, and so on. The coney garth was not
a mere name, but was well stocked with game, and even
at a much later date, from Edward IV. to Henry VIII.,
there were various ordinances in force for punishing law
students who hunted rabbits with bows and arrows or
darts.

In the first year of Queen Elizabeth the Garden was
separated from the fields by a clay embankment, and a
little later a brick wall was added, with a gate into the
fields, which is probably the same as the present little
gate to the north of the new hall, at the end of the
border, shown in the illustration. The Garden continued
much further along the wall then, and only was curtailed
when the new hall and library were built in 1843.
The delightful terrace which is raised against the wall
overlooking the "fields" was made in 1663. On June
27th of that year, Pepys, who on other occasions

LINCOLN'S INN

mentions his walks there with his wife, went to see the alterations. "So to Lincoln's Inne, and there walked up and down to see the new garden which they are making, and will be very pretty." The outside world seems to have had easy access to the gardens of all the Inns of Court in those days, but it was regarded as a special privilege granted to a very wide circle, and a favour not accorded to the public at large. In the *Tatler* occur such passages as, "I went into Lincoln's Inn walks, and having taken a round or two I sat down according to the allowed familiarity of these places." Again, "I was last week taking a solitary walk in the garden of Lincoln's Inn, a favour that is indulged me by several of the benchers who are my intimate friends."

They were, however, so much frequented by all the fashionable world of London, that the foreigner arriving there naturally took them for public gardens. Mr. Grosley, who came to London in 1765, thus describes them :—

"Besides St. James's Park, the Green Park, and Hyde Park, the two last of which are continuations of the first, which, like the Tuileries at Paris, lie at the extremity of the metropolis, London has several public walks, which are much more agreeable to the English, as they are less frequented and more solitary than the Park. Such are the gardens contained within the compass of the Temple, of Gray's Inn and Lincoln's Inn. They consist of grass plots, which are kept in excellent order, and planted with trees, either cut regularly, or with high stocks : some of them have a part laid out for culinary uses. The grass plots of the gardens at Lincoln's Inn are adorned with statues, which, taken all together, form a scene very pleasing to the eye."

The students must certainly have aimed at keeping their gardens from the vulgar gaze, and showed their displeasure at some one who had built a house with windows overlooking the Garden in 1632 in an uproarious manner. They flung brickbats at the offending window until "one out of the house discharged haile shot upon Mr. Attornie's sonne's face, which though by good chance it missed his eyes yet it pitifully mangled his visage."

Old maps of the gardens show a wall dividing the large upper garden from the smaller, but by 1772 the partition had disappeared. It was doubtless unnecessary when the terrace was made and the rabbits done away with.

The 1658 map with the wall in it shows the upper garden intersected by four paths, and an avenue of trees round three sides, and the small garden with a single row of trees round it divided into two large grass plots. The lovely shady avenue below the terrace in the large garden has still a great charm, and although not so extensive as it once was, the great green-sward and walks seem very spacious in these days of crowding. The terrace overlooking Lincoln's Inn Fields, with the broad walk and border of suitable old-fashioned herbaceous plants, has great attractions. The view from here must have improved since the days when the Fields were a wild-looking place of evil repute, and the scene of bloody executions. In the lonely darkness below the terrace wall, deeds of violence were only too common.

> "Though thou are tempted by the linkman's call,
> Yet trust him not along the lonely wall.
> In the mid-way he'll quench the flaming brand,
> And share the booty with the pilfering band."
> —GAY.

Certainly when one is sentimental over the departed charms of Old London, it would be an excellent antidote to call up some of the inconveniences that electric light and the metropolitan police have banished.

There is more character about the gardens of Gray's Inn than either the Temple or Lincoln's Inn. They have come down with but little alteration from the hands of that great lover of gardens, Bacon. But long before his time gardens existed. The land on which Gray's Inn stands formed part of a prebend of St. Paul's of the manor of Portpoole, and subsequently belonged to the family of Grey de Wilton, and in the fourteenth century the Inn of Court was established. Between its grounds and the villages of Highgate and Hampstead was an unbroken stretch of open country. There, in Mary's reign, Henry Lord Berkeley used daily to hunt "in Gray's Inne fields and in those parts towards Islington and Heygate with his hounds," and in his company were "many gentlemen of the Innes of Court and others of lower condition . . . and 150 servants in livery that daily attended him in their tawny coates." In Bacon's time it must still have been as open, and Theobald's Road a country lane with hedgerows. The Garden already boasted of fine trees, and among the records of the Society there is a list of the elms in 1583 all carefully enumerated, and the exact places they were growing: "In the grene Courte xi Elmes and iii Walnut trees," and so on. Eighty-seven elms, besides four young elms and one young ash, appear on the list; so the Garden was well furnished with trees even before Bacon commenced his work. Gray's Inn was the most popular of the four Inns of Court in the Elizabethan period, and many famous men, such as Lord Burghley, belonged to it. It was in 1597 that

Bacon took the Garden in hand, some ten years after he became a Bencher. In the accounts of that year £7. 15s. 4d. appears "due to Mr. Bacon for planting of trees in the walkes." In 1598 it was resolved to "supply more yonge elme trees in the places of such as are decayed, and that a new Rayle and quicksett hedge be sett upon the upper long walke at the good discretion of Mr. Bacon, and Mr. Wilbraham, soe that the charges thereof doe not exceed the sum of seventy pounds." On 29th April 1600, £60. 6s. 8d. was paid to "Mr. Bacon for money disbursed about garnishing of the walkes."

Bacon's own ideas of what a garden should be are so delightfully set forth in his essay on gardens, that the whole as it left his hand is not difficult to imagine. The fair alleys, the great hedge, were essentials, and the green, "because nothing is more pleasant to the eye than green grass kept finely shorn." His list of plants which bloom in all the months of the year was compiled of those specially suited "for the Climate of London," so no doubt some would be included in this Garden under his eye, although they do not appear in the records. He wished "also in the very middle a fair mount," and even this desire he carried out in Gray's Inn. In a description of the Garden as late as 1761, a summer-house which Bacon put up in 1609 to the memory of his friend Jeremiah Bettenham is mentioned as only recently destroyed. "Till lately," it says, "there was a summer-house erected by the great Sir Francis Bacon upon a small mount: it was open on all sides, and the roof supported by slender pillars. A few years ago the uninterrupted prospect of the neighbouring fields, as far as the hills of Highgate and Hampstead, was obstructed by a handsome row of houses on the north; since which the above summer-house has been levelled, and many trees cut down to lay the Garden more open." The view, even then, was fairly open, as Sir Samuel Romilly, in 1780, complains of the cold, as there was "only one row of houses" between him and Hampstead, and "a north-west wind blows full against" his chambers. This "most gallant prospect into the country, and its beautiful walks" were the great attractions of these Gardens. They appear to have been one of the most fashionable walks, especially on Sundays. Pepys was frequently there, and his diary records, several times, that he went to morning church, then had dinner, then to church again, and after went for a walk in Gray's Inn. That he met there "great store of gallants," or "saw many beauties," is the usual comment after a visit. On one occasion, he took his wife there to "observe the fashions of the ladies," because she was "making some clothes." The walks and trees are redolent with associations, and the Gardens, though curtailed, have much the same appearance as of yore. When a portion of the ground was sacrificed to the new buildings, those who loved the Garden deeply bewailed. "Those accursed Verulam Buildings," wrote Charles Lamb, recalling his early walks in Gray's Inn Gardens, "had not encroached upon all the east side of them, cutting out delicate green crankles, and shouldering away one of two stately alcoves of the terrace. The survivor stands gaping and relationless, as if it remembered its brother. They are still the best gardens of any of the Inns of Court—my beloved Temple not forgotten — have the gravest character, their aspect being altogether reserved and law-breathing.

Bacon has left the impress of his foot upon their gravel walks."

After such a delightful summary of their charms it seems cruel to try and dispel one of their most treasured traditions—namely, that Bacon planted the catalpa.   It is a splendid and venerable tree, and there is no wish to pull it from its proud position of the first catalpa planted, and the finest in existence in this country ; but it is hard to believe that Bacon planted it, in the light of the history of the plant. There is no mention of a catalpa in any of the earlier writers—Gerard did not know it, and it is not in the later edition of his work by Thomas Johnson, in 1633, or in Parkinson's "Paradisus," in 1629, or in Evelyn's "Sylva," in 1664, all published after Bacon's death.

The tree was first described by Catesby in his "Natural History of Carolina," a splendid folio which appeared in 1731.   There it is classed as *Bignonia urucu foliis*, or *Catalpa*, as it was not until later that Jussieu separated the genus *Catalpa*.   He says the tree was not known to the inhabitants of Carolina till the seeds "were brought there from the remoter parts of the country," "and though the inhabitants are little curious in gardening, the uncommon beauty of this tree induced them to propagate it, and it is become an ornament to many of their gardens, and probably will be the same to ours in England, it being as hardy as most of our American plants :   many of them, now at Mr. Bacon's, at Hoxton, having stood out several winters without any protection, except the first year."   Hoxton was then a place famous for its nursery gardens.   In 1767, in Catesby's volume on the trees of North America, he gives the same story, and adds, "in August 1748" it produced, "at Mr. Gray's, such numbers of blossoms, that the leaves were almost hid thereby." This Mr. Gray owned the nurseries in Brompton, famous under the management of London and Wise.

In Philip Miller's dictionary, Catesby's history of the plant is referred to, and also in 1808, in the *Botanical Magazine*, when the plant was figured. There it says the plant "has been long an inhabitant of our gardens, being introduced by the same Botanist [Catesby] about the year 1728." "It bears the smoke of large towns better than most trees; the largest specimen we have ever seen grows in the garden belonging to the Society of Gray's Inn." There is no hint that the tree in question could have been here before Catesby's discovery, and it is not till Loudon's Encyclopædia in 1822 that the planting is attributed to Bacon. Such a remarkable tree could hardly have escaped all gardeners for more than a century, during a time when gardening was greatly in fashion, and every new plant greedily sought after. We know that nearly a hundred years ago this specimen was the finest in England, and therefore it may have been planted not more than a hundred years or so after Bacon's death. Raleigh very likely walked with Bacon on the spot where it now stands, but, alas! the possibility that he brought Bacon a tree from Virginia, which was only discovered near the Mississippi a century later, is hardly credible.

The entrance to the Gardens on the Holborn side is through massive wrought-iron gates, on which the date 1723 is legible. The letters "w. i. g." are the initials of the Treasurer during whose tenure of office they were erected, the "T" above standing for Treasurer. In the Inns of Chancery a "P" for Principal, associated with the various initials, is often to be noticed. These fine gates are a charming approach to the sequestered walks and ancient trees. Gray's Inn Gardens have another delightful speciality, in that the rooks delight to honour them by building there. They have a warm welcome, and good food in cold weather, and seem likely to remain. Looking through the lofty iron gates, the rooks' nests are seen, and the pleasant cawing sound adds greatly to the attraction of the place.

THE GARDEN GATES, GRAY'S INN

CHAPTER XII

HISTORICAL GARDENS

History is philosophy teaching by examples.
—BOLINGBROKE.

LTHOUGH their number has sadly diminished of late years, London still has a few spaces remaining which may be classed as gardens. Often they are merely green patches of a formal type, which are better suited to the present climate than attempts at flowers; but a few regular gardens still exist, bringing dreams of a former period. In St. Bartholomew's Hospital, the oldest of all such institutions, the square, with a handsome fountain in the centre, is more what one expects to find in Italy than in Smithfield. It is this sort of surprise that makes the charm of London, and renders a wander through its mazes so attractive. What a contrast the walk of a few minutes can bring in the heart of London! but of all these changes none is more impressive than the hush of the Charterhouse after the rush of Aldermanbury or the noise of Clerkenwell. There is still lingering there the touch of the old monastery; a breath of a bygone age seems to pervade the courtyards and gateways, and something in

the silence speaks of another world. The first indica-
tion of its hidden green courts are the mulberry leaves
peeping over the worn stone wall, near the gateway
which leads to the weathered archway, the entrance of
the old Carthusian monastery. This is the very spot
where, with the brutal severity of Tudor times, the
arm of the last Prior was exposed after his cruel
execution at Tyburn. The monastery, founded in 1371,
was dissolved with unusual barbarity, and passed into
secular hands. The possession of it by the Duke of
Norfolk has left its mark in many of the existing
buildings, as he converted it from a cloister to a palace,
but its palatial days did not last long. It was bought
by the benevolent Thomas Sutton, a portion of whose
large fortune, amassed from profitably working coal
mines, was bestowed in founding "a hospital for poor
brethren and scholars." The scholars have been taken
away from the historical associations, to the purer air
of Godalming, and the parts of the buildings devoted
to their accommodation were in 1872 bought by the
Merchants Taylors' Company for their school. The
playing field of the boys is the ample space which was
enclosed by the cloister of the monastery. Part of the
land to the north has been built over, and a tall ware-
house overlooks the burying-ground of the monks,
which is still a large green sward of hallowed ground,
with a row of mulberries. This lies so far below the
level of Clerkenwell Road that a flight of steps leads
to the postern gate in the high wall, overhung with
climbing plants. This "God's acre" is covered with
smooth turf, and some day the two walnut trees planted
by the master in 1901 may afford grateful shade. It
is in keeping with the spirit of the place to plant trees

of such slow and stately growth. The Preachers' Court and the smaller Pensioners' Court are like college quadrangles, with that perfect turf that England alone produces. The smooth surface is broken only by the regular intersecting gravel paths, and one row of mulberry trees some seventy years old. The red-brick buildings have a venerable appearance, although they do not carry the weight of centuries with dignity, like the " Wash-house Court," the hall, the library, or the brick cloister and the delightful old walls with their deliciously-scented fig-trees. The whole place has a mediæval look and feeling, and teems with ghosts and recollections of the monks of the early peaceful days, and their courageous successors at the Dissolution. The pious founder, as the chorus of the old Carthusian melody says, must not be forgotten :—

" Then blessed be the memory
Of good old Thomas Sutton,
Who gave us lodging, learning,
As well as beef and mutton."

Of the shades which surround these peaceful green courts none appear more real than that of Colonel Newcome. The guardian will point out the room in which he died, or his pew in the chapel, as if he belonged to history as much as Wray, who bequeathed the old books in the " Officers' Library," or any of the well-known pensioners. With such true and pathetic touches has Thackeray drawn the character of Colonel Newcome that fiction has here become entwined round the walls almost as closely as fact.

Further eastward is an open piece of ground, which is hardly a garden ; but as it is green, and took the place of what was known as the Artillery Garden, it may claim a

moment's consideration. Push open a door in the
modern-looking castellated building in the City Road
near Bunhill Fields, and a large, quiet, open space is
discovered. Old guns look inoffensively down on a
wide square of green turf. This is the home of the
Honourable Artillery Company, the descendants of the
"Trained Bands" of citizens, first enrolled in 1585 in
the fear of a Spanish invasion. They have been here
since 1622, when they moved from near Bishopsgate
Without. "Artillery Garden," or Teazel Close or
Garden, was the name of the older place, from the teazel
grown there for the cloth workers.

> " Teazel of ground we enlarge St. Mary's Spittle,
> Trees cut down, and gardens added to it,
> Thanks to the lords that gave us leave to do it,"

says an old poem. The existing Artillery Ground was
a great place for cricket matches, where county met
county in the eighteenth century. It was here that a
vast crowd witnessed the first balloon ever launched into
the air in England, sent up by Count Zambeccari in
1783. The next year, from the same place, Lunardi was
more ambitious, and actually went up in his balloon.
It proved too small for the friend who was ready to
risk his life in his company, so he took a dog, a cat, and
a pigeon with him instead.

Passing on into the City, the remains of the once
extensive Drapers' Garden is met with.[1] Only a small
piece, seen from the street through iron railings, and
approached through the hall, has been retained ; a few
trees and bright flowers survive of what was once a
fashionable and much sought after resort.

[1] See page 12.

Most of the other patches of green in the City are
disused burial-grounds, and are considered in a chapter
by themselves. Beyond the City, on the east, in the Mile
End Road, is the quiet old Trinity Hospital. It stands
on the north of that wide road, which might be made
one of the most beautiful entrances to the City. The

TRINITY ALMSHOUSES, MILE END ROAD

simple good taste of these delightful old almshouses is
a great contrast to some of the surroundings. They
were probably designed by John Evelyn, with the assistance of Wren. His father-in-law, Sir Richard Browne,
founded and built very similar almshouses at Deptford,
long since swept away. Of these Evelyn writes, "It
was a good and charitable work and gift, but would have
been better bestowed on the poor of that parish than on
seamen's widows, the Trinity Company being very rich,

and the rest of the poor of the parish exceedingly indigent." In spite of these sentiments, he is believed to have had a hand in the Mile End Almshouses, which were founded by Captain Henry Mudd of Ratcliffe, Captain Sandes or Sanders, and Captain Maples. The two last are remembered by statues still standing in the little formal gardens. Maples, who appears in the dress of a naval officer of the period, left a fortune for the use of the guild in diamonds, collected in India, where he was an early pioneer, and where he died in 1680. A similar endowment in Hull is described in a poem in 1662 :—

> " It is a comely built, well-ordered place,
> But that which most of all the house doth grace
> Are rooms for widowes, who are old and poore,
> And have been wives to mariners before."

Certainly Trinity Hospital, Mile End, is comely and well ordered. The pensioners take a pride in keeping every nook and corner scrupulously clean. Everything is, in fact, in "ship-shape" order. The grass is neatly mown, the trees on either side well trimmed and clipped. Outside each little house a few plants are carefully tended, the pots arranged with precision, and every flower looked after with pride. It is indeed a peaceful place for these old people to pass their declining years in, and the sight makes the regret for St. Katharine's and the other vanished charitable buildings all the more keen.

The site of another benevolent institution near is fulfilling a useful and delightful task, although the old houses attached to it have disappeared. It was a row of almshouses founded by a member of the Brewers' Company, named Baker, about 150 years ago, for widows.   The garden was much too large for these decrepid old women to cultivate, so the place was taken in hand some twenty-five years ago by the Rev. Sidney Vatcher, who built the beautiful church of St. Philip, Stepney, hard by, and he became the tenant of the Brewers' Company.   This charming garden was at first more or less opened by him to the parish, but lately it has been put to the most suitable use of giving a quiet place for rest and recreation to the nurses of the London Hospital.   The almshouses were pulled down about four years ago, to make way for the laundries of the Hospital.   Here, indeed, is one of those sudden and surprising contrasts to be found in London.   A high brick wall encloses this oasis, and the nurses and some privileged people have keys to the door, which opens, from a side street close to the noise of the Mile End Road, suddenly into a peaceful, picturesque garden.   The idea in the formation was a willow-pattern plate, and the little bridge over a miniature stream is reproduced. Plane trees in a formal array are kept trimmed to give a dense shade, and the hammocks hung from them in summer provide the most ideal resting-places for the worn-out nurses.   At one time animals were kept here in cages, as a kind of small "Zoo" for Whitechapel; but since the last alterations the animals have been relinquished, and the bear-pit makes a delightful rock garden, and the various other cages form summer-houses. One thoughtful addition of the vicar was placing a small stove in one of these shelters, with an array of kettles, teapots, cups and saucers, so that any of the nurses resting can have their *al fresco* cup of tea—and what could be more grateful and comforting?   A French writer who recently gave her impressions of L'Ile Inconnue was charmed with the peace and repose of this little East End Paradise. After seeing the Hospital and all its wonderful appliances, "You will now see our Eden," said the guide. "Ici! l'Eden! m'écriai-je, après le péché alors!" Then, when she had for a moment looked within those mysterious high walls, "N'avais-je pas raison d'appeler ce jardin l'Eden?" said the friend. "Oui, repondis-je, c'est l'Eden après la Rédemption." Certainly any one who sees this little garden, and realises the devoted lives of those who made it and those who enjoy it, must agree with this writer.

It is not often that, when the old almshouses vanish, the neighbourhood benefits to such an extent. What will be the fate of the Ironmongers' Almshouses in Kingsland Road, between Shoreditch and Dalston? A large board in the garden that fronts the street announces the site is for sale!

The Foundling Hospital has large green courts, on which the merry but sombrely-clad little children are seen running about, through the fine iron gates which face Guildford Street. This was founded in 1739 by Captain Thomas Coram, who gave so much of his wealth to objects of charity and philanthropy that a subscription had to be raised to support him in his old age. Theodore Jacobson (died 1772) was the architect of the building. A colonnade runs round the whole length of the forecourt up to the gates, part of which is used as laundries, or other things necessary to the institution. A writer in 1773 describes the "large area between the gates and the hospital" as "adorned with grass plats, gravel walks, and lamps erected upon handsome posts: beside which there are two convenient gardens," and exactly the same description holds good to-day. Brunswick Square lies to the west, and Mecklenburgh Square to the east, so the Hospital grounds are still airy. There is a small garden at the back of the building in front of the Infirmary; on the east is the Treasurer's Garden, a fair-sized enclosure, and on the other side, with the poplars growing in Brunswick Square overhanging it, lies the other and larger of the two "convenient gardens." There is nothing old-fashioned or attractive in these gardens left; merely a green lawn, a weeping ash, and a few commonplace "bedding-out" plants; not altogether in keeping with the age or dignity of the building and spacious forecourt.

Less well known is the delightful Garden of the Grey-coat School in Westminster. Most of the old foundations in Westminster have vanished, such as Emanuel Hospital and the "Blue-coat School," which disappeared a few years ago, but so far this charming old house has been respected. Quaint figures of the children in the dress of the time—it was founded by the citizens of Westminster in 1698 — stand on either side of the entrance. The children from the parishes of St. Margaret and St. John the Evangelist, who have attended the elementary schools for three years, are eligible for admission, up to the age of ten. The school was reconstituted as a day school for 300 girls in 1873, and, in spite of all educational vicissitudes, has been allowed to survive, and the sweet and wholesome influence of those old-fashioned surroundings would be a great loss, should it ever be swept away. The Garden is delightful. It is practical as well as ornamental, as it furnishes the staff of teachers with a good supply of vegetables. They have each a small flower-bed too, tended with great care, and the children are allowed a

place of their own, where they work, dig, and plant.
Down the centre runs a wide gravel walk, with a deep
herbaceous border along either side, sweet-scented pinks
and low-growing plants near the front, then a long row
of spiderwort, and behind that a regiment of magnificent
hollyhocks. The spiderwort or Tradescantia is a flower
eminently suited to London gardens, not only because it
seems to withstand any amount of smoke and bad air,
but because of its association with the famous garden in
Lambeth, where it was first grown. Parkinson, in 1629,
gives the history of his friend's introduction of the plant.
"The Spiderwort," he writes, "is of late knowledge, and
for it the Christian World is indebted unto that painfull
industrious searcher, and lover of all nature's varieties,
John Tradescant (sometimes belonging to the Right
Honourable Lord Robert Earle of Salisbury, Lord
Treasurer of England in his time, and unto the Right
Honourable the Lord Wotton at Canterbury in Kent,
and lastly unto the late Duke of Buckingham), who first
received it of a friend, that brought it out of Virginia,
thinking it to bee the Silke Grasse that groweth there,
and hath imparted hereof, as of many other things,
both to me and others." "Unto this plant I confess I
first imposed the name . . . which untill some can finde
a more proper, I desire may still continue . . . John
Tradescant's Spider Wort of Virginia." Courageous as
herbalists generally were in tasting plants, Parkinson
confesses there had "not beene any tryall made of the
properties" or "vertues." Luckily no one has disputed Parkinson's choice of a name, and his friend's
memory is still preserved. The plant is not confined
to Virginia, but grows much further into the Wild West,
and is common in Kansas, Nebraska, and distant States.

GREY COAT SCHOOL, WESTMINSTER

Yet it will still adapt itself to the grimy limits of a London garden, and flower year after year. The Greycoat School Garden is quite refreshing; the plants look so healthy and prosperous that it is really encouraging. The interior of the house, with oak beams and panels, is all in keeping, and the long class-room, with windows looking out on the bright Garden, is most ideal. As, at the close of their afternoon studies, the girls, singing sweetly in parts, join in some familiar hymn, and the melodious sounds are wafted across the sunlit Garden, it is hard to believe in the existence of the crowded, unsavoury slums of Westminster, only a stone's throw from this "haunt of ancient peace."

Among its many charms and associations Westminster Abbey can lay claim to possessing one of the oldest gardens in England. The ground still occupied by the space known as the "College Garden" was part of the infirmary garden of the ancient monastery. It cannot trace back its history with the Abbey to the Saxon Sebert, but when Edward the Confessor's pile began to rise, and all the usual adjuncts of a monastery gathered round it, the infirmary with the necessary herb-garden of simples for treating the sick monks would be one of the first buildings to be completed. One of the most peaceful and retired spots within the Abbey precincts is the Little Cloister, which was the infirmary in early days. When the Great Cloister was finished in 1365, the Little Cloister was taken in hand. Payments for work on "the New Cloister of the Infirmary" appear in the accounts from 1377, and it was completed in 1390, and that year the centre was laid down in turf. The garden belonging to the infirmary covered all the space now occupied by the "College Garden," and joined the

"Grete Garden," which lay to the west. It was probably, like all the gardens of that date, laid out in long, narrow, straight beds, in which were grown all the healing herbs used for the sick of the monastery. Probably there were fruit-trees, too, as in 1362 John de Mordon, the infirmarer, got 9s. for his apples, and the following year 10s. for pears and apples. No doubt the favourite Wardon pear was among them, as in another record, between 1380–90, it is specially mentioned. The chapel of St. Katharine, which stood on the north side of the Garden, was destroyed in Elizabeth's reign. This, the infirmary chapel of Norman building, was as replete with history as every other nook and corner of the Abbey buildings. Here St. Hugh of Lincoln and most of the early bishops were consecrated, and here took place the unseemly dispute for precedence, between the Primates of Canterbury and York in 1186, which led to the settling of their respective ranks by the Pope. While so many changes have swept over the Abbey, and whole buildings have vanished, the herb-garden of early days has kept its place, and is still a garden, though bereft of its neat little beds.

The Little Cloister has been greatly altered since then, having been refashioned in the early part of the eighteenth century under the influence of Wren. Although so changed since the time when strange decoctions of medicinal herbs were administered within its walls, it has retained much of its fascination, and the approach to it by the dim vaulted entrance, dating from the Confessor's time, out of the narrow passage known as the " Dark Entry," adds to its charm. The sun streams down on this small court, with its tree and ferns and old moss-grown fountain, lighting it with a kind of "dusky splendour."

Victoria Clemente.
1906

ABBEY GARDEN, WESTMINSTER

Any one standing in this suggestive spot will feel with Washington Irving, that "The Cloisters still retain something of the quiet and seclusion of former days. The gray walls are discoloured by damps, and crumbling with age; a coat of hoary moss has gathered over the inscriptions of the mural monuments, and obscured the death's heads, and other mural emblems. The sharp touches of the chisel are gone from the rich tracery of the arches; the roses which adorned the keystones have lost their leafy beauty; everything bears marks of the gradual dilapidations of time, which yet has something touching and pleasing in its very decay."

These lines refer to the Great Cloister, but the quiet and repose are still more noticeable in the Little Cloister, which rarely echoes to the sound of hurrying feet. The noise and laughter of Westminster scholars is only dimly heard in this secluded corner. The boys are not as boisterous as when Horace Walpole feared to face them alone, even to visit his mother's tomb. "I literally had not courage to venture alone among the Westminster boys; they are as formidable to me as the ship carpenters at Portsmouth," he wrote in 1754. Even in those days the list of eminent scholars was already a long one— Hakluyt, Ben Jonson, George Herbert, Dryden, Wren, being on the roll of those who had passed away, besides others then living, such as Gibbon and Warren Hastings, who carried on the tradition of this classic ground.

In monastic times there were many gardens within the precincts of the Abbey, besides the infirmary garden; but it is difficult to locate all of them with certainty, although the sites of some are known. The abbot's garden lay in the north-west angle of the wall, and must have covered part of the present Broad Sanctuary, including

THE LITTLE CLOISTER, WESTMINSTER ABBEY

the spot where the Crimean monument now stands. Beyond the abbot's house, just west of the cloister, was the abbot's little garden. The northern part of Dean's Yard was from very early times known as "The Elms," from the grove of fine trees, some of which remain. It is said that when Elizabeth ascended the throne and summoned Abbot Feckenham, who had been reinstated by Mary, he was planting some, perhaps these identical, elm trees. Among them formerly stood a huge oak, which was blown down in 1791. The horse pool was on the west of the Elms, and beyond both to the south lay the numerous adjuncts of the monastery, the brewhouse, bakehouse, and granaries. Skirting this enclosure was the "Long Ditch," which flowed by the line of the present Delahay Street and Prince's Streets, and passed along outside of the wall of the Infirmary Garden, in what is now Great College Street, and fell into the Thames. This stream turned the mill from which "Millbank" took its name. In it, to the south of the granary, was a small island osier bed. The sale of the osiers on it used to bring in 10s. annually in the fourteenth century. Beyond the stream were more gardens. The "Hostry Garden" was a large one on the site of the church of St. John, and next to it the "Bowling Alley," where Bowling Street ran in later times, and to the west of that was a kitchen-garden. Somewhere also on the west of the "Long Ditch," before it turned towards the Thames near the osier island, must have been the "Precentor's Mede," or, as it was sometimes called, the "Chaunter's-hull," and also the "Almoner's Mede" or "Almery Garden." On the other side of the "Hostry Garden," southwards on the site of "Vine Street" and "Market Street," was situated the vineyard, without which no

thirteenth-century monastery was complete, and "Market
Mede." Even this does not exhaust the list of separate
gardens, but the others probably lay further away. The
cellarer had charge of a large garden, which may have been
the "Convent Garden," which is so familiar as "Covent
Garden" that the connection between the site of the
market and the Abbey has been lost sight of. One
of the large gardens which was generally let was
"Maudit's Garden." In the records it is spoken of as
"Maudit's" or "Caleys." The name Maudit was given
to it because Thomas Maudit, Earl of Warwick, in the
thirteenth century effected an exchange of lands with
the Abbey, of which the garden formed a part. The
other name, "Caleys," was "Calais," named from the
wool staplers who came from that town and resided
near there, just as "Petty France" (where Milton lived)
was called so from the French merchants. An Act of
interchange of land between Henry VIII. and the Abbey,
in the twenty-third year of his reign, mentions "a
certain great messuage or tenement commonly called
Pety Caleys, and all messuages, houses, barns, stables,
dove-houses, orchards, gardens, pools, fisheries, waters,
ditches, lands, meadows, and pastures." Part of this was
"Maudit's" garden, which was sometimes in the hands
of the convent, but more frequently let out. Among
the muniments in 1350, "a 'toft called Maudit's garden,
and a croft called Maudit's croft," are referred to. There
seems to have been an enclosure within this "toft" which
was let out separately, and in the twentieth year of
Edward IV., Matilda, the widow of Richard Willy, who
had held it, gave up this enclosure or "conyn garth."
This was probably a "coney garth" or rabbit enclosure,
like the one at Lincoln's Inn, which was kept up for

a long time.   Such rabbit gardens were by no means uncommon.   All gardening operations must at times have been rendered difficult by reason of the wet soil and frequent flooding of the river, but with the patient persistence characteristic of gardeners in those days, the gardens in monastic times were probably well kept, and yielded profitable crops.   It is delightful to know that, in spite of all the changes, one portion of the old gardens actually remains to this day.

Lambeth, on the opposite bank, fared no better than Westminster for high tides, and wet seasons did occasional damage there.   In Archbishop Laud's Diary, he notes the inroad of a high tide, which certainly would be destructive :—"November 15, 1635, Sunday.   At afternoon the greatest tide that hath been seen.   It came within my gates, walks, cloysters, and stables at Lambeth."   Nothing of great antiquity now remains in these Lambeth Gardens, although they are indeed historic ground.   The long terrace and wide herbaceous border, with a profusion of madonna lilies, backed by a wooden paling, and fruit-trees peeping over, is now a charming walk.   The trees on the right of the illustration are planes, ailanthus, and catalpas, all smoke-resisting and suitable, but not such as would have ornamented the Garden in older days, when Archbishop Cranmer adorned his garden with "a summer-house of exquisite workmanship."   It was designed by his chaplain, Dr. John Ponet or Poynet, who is said to have had "great skill and taste in works of that kind."   The summer-house was repaired by Archbishop Parker, but afterwards fell into decay and was removed, and in 1828 not even a tradition of where it had stood remained. The site of " Clarendon's Walk," another historical

U

corner of the Lambeth Garden, is also uncertain. It appears to have received the name from a conversation which took place in the Garden between Laud and Hyde, in which the latter seems to have told the Archbishop pretty plainly that "people were universally discontented . . . and many people spoke extreme ill of his grace," on account of his discourteous manners, which culminated on one occasion by his telling a guest "he had no time for compliments," which greatly incensed him. The only survivals of former years are the delightfully fragrant fig-trees, which flourish between the buttresses on the sunny side of the library—the great hall rebuilt by Archbishop Juxon after the destruction in Cromwell's time. These figs are now fair-sized trees, but they are only cuttings of the older ones destroyed in 1829, when Archbishop Howley commenced his rebuilding. The two parent trees, in 1792, measured 28 inches and 21 inches in circumference, and were 50 feet high and 40 feet in breadth, and, according to contemporary evidence, bore delicious fruit of the white Marseilles variety. Tradition ascribed their planting to Cardinal Pole during his brief sojourn as Archbishop.

Latimer seems much to have appreciated the Lambeth Garden, when business called him to the Palace. Sir Thomas More describes, in 1534, how he watched him walking in the Garden from the windows. Latimer himself, in writing to Edward VI., says, "I trouble my Lord of Canterbury, and being at his house now and then, I walk in the Garden looking at my book, as I can do but little good at it. But something I must needs do to satisfy the place. I am no sooner in the Garden and have read awhile, but by-and-by cometh

HERBACEOUS BORDER, LAMBETH PALACE

there some one or other knocking at the gate. Anon cometh my man and saith, 'Sir, there is one at the gate would speak with you.'" How many of us that have been called in from a pleasant garden to perform some unpleasant task will sympathise with the Bishop!

One famous inhabitant of the Garden lived through many and great changes. This was a tortoise, which is said to have been put into the Garden by Archbishop Laud, and lived until 1757, when he perished by the negligence of a gardener. This legend is apparently quite true, so it had been there for over 110 years.

A short account of the principal gardens near London, written by Gibson in 1691, describes that of Lambeth Palace. It "has," he says, "little in it but walks, the late Archbishop [Sancroft] not delighting in" gardens, "but they are now making them better; and they have already made a green-house, one of the finest and costliest about the town. It is of three rooms, the middle having a stove under it; . . . but it is placed so near Lambeth Church, that the sun shines most on it in winter after eleven o'clock, a fault owned by the gardener, but not thought of by the contrivers. Most of the greens are oranges and lemons, which have very large ripe fruit on them." The Archbishop who thus took the garden in hand was Tillotson, and it is not surprising to find him adopting that keenness for gardening and the cultivation of "greens" brought into fashion by William III.

Nearly ten acres of the extensive grounds of Lambeth Palace have now been put under the management of the London County Council, and made open to the public as "Archbishop's Park." For many years this Park had been used for cricket and so on, but the transference entailed some alterations, and extended its use to a wider circle.

The Garden of Fulham, the other ecclesiastical palace of London, is even more interesting than Lambeth, on account of the fine trees still remaining of which the history is known. Among the Bishops of London several have shown great interest in the gardens, and two especially, Grindal and Compton, were eminent gardeners. The tamarisk was introduced by Bishop Grindal, and in the golden age of gardening he was in the foremost rank of the patrons of the art, with Bacon and Burghley. He used to send Queen Elizabeth presents of choice fruits from his garden, and on one occasion got into trouble by sending fruit, when one of his servants was supposed, unjustly, to have the plague. He wrote (5th August 1566) to Burghley, to say he was sorry he had "no fruit to offer him but some grapes." These grapes were of course produced out of doors, as growing vines in green-houses was a fashion unknown until some 150 years later. Even before the additions of Grindal, the gardens were extensive, and Bonner is said to have been much in his garden, not from the love of its repose, but, according to contemporary but prejudiced chroniclers, because in the further arbours of the garden he could with the rod or by other equally stringent measures, "persuade" undisturbed those of the reformed religion to recant and adopt his views. His successor, Grindal, used the Garden for more laudable and peaceful practices, and his work of planting was much appreciated in that garden-loving age. Bishop Aylmer, who, after Sandys, succeeded Grindal in 1577, was accused of destroying much of Grindal's work and cutting down his trees, then some thirty-five years old. Strype, however, protests that he only cut down " two or three of the decayed ones." That there should be a controversy on the subject only shows how much was thought of Grindal's planting. The same thing happened after the death of Compton, the next great planter, as Robinson, who followed him, let the gardener sell and cut down as much as he liked. In our own day, even, some of Compton's elms have been removed, to make the alterations in the Bishop's Park when it was opened to the public. The Bishop's Park is the long, narrow strip of land between the moat and the river. Flowering shrubs on the bank of the moat, and rows of cut plane trees by the river, have been planted. There are two long asphalt paths, and some bedding out and rock gardening between the grass lawns. It is now kept in order by the Borough of Fulham, which reminds the public of the fact by the notices stuck up: " Ratepayers, protect your property."

The Elm Avenue was part of Compton's design, and many very fine trees known to be his remain to this day. During the long duration of his episcopate—1675 to 1714—he had time to see his plants grow and flourish. His gardening achievements were much appreciated in his own day. John Evelyn, a great authority on horticultural matters, was often at Fulham. He notes in his Diary on Oct. 11, 1681 : " To Fulham to visit the Bishop of London, in whose garden I first saw the *Sedum arborescens* in flower, which was exceedingly beautiful." Richard Bradley, a well-known gardener, in his book published in 1717, quotes many of the plants at Fulham as examples in his pages. With regard to the passion flower, his notice is interesting, as it gives the name of Bishop Compton's gardener. "That [the passion flower] may bear fruit," he writes, "we must Plant it in very moist and cool places, where it may be continually fed with Water; this I had from the Curious Mr. Adam Holt, Gardener to the late Bishop of London, who shew'd me a letter from the West Indies, from whence I learnt it was an Inhabitant of Swampy Places." Bradley had seen the pistachio fruiting against a wall at Fulham, and he thought he had also noticed an olive flourishing there. From time to time there have been special notices of the trees round the Bishop's palace. Sir William Watson wrote a paper on them for the Royal Society, in which he gives a list of thirty-seven special trees, many of them the finest of their kind in England. "For exemplification of this I would," he says, "recommend to the curious observer the black Virginian walnut tree, the cluster pine, the honey locust, the pseudo-acacia, the ash maple, &c., now remaining at Fulham." Many of the later bishops have paid great attention to the grounds. Bishop Porteous (1787–1809) who planted cedars; Howley (1813–1828), and especially Blomfield (1828–1856), all took delight in the Garden. Bishop Blomfield planted a deciduous cypress and the ailanthus, which now measures 10 feet 4 inches at 4 feet from the ground, curiously exactly the same girth as the one at Broom House close by. In 1865, Bishop Tait had the old trees measured, and there are later measurements of some of the finest. The cork tree was 13 feet 9 inches, and although sadly shattered, part of this magnificent old tree, with its thick cork bark, still holds its own. The great black walnut or hickory has not been so fortunate, and died about ten years ago, and only a venerable stump is left; but a good specimen still stands in the meadow. The great tree in 1865 measured 15 feet 5 inches; in 1894, 17 feet 3 inches. The tulip tree died about the same time as the hickory. The honey locust (*Gleditschia triacanthos*), one of Bishop Compton's trees, only died last year, the large white elm in 1904, and, sad to say, the flowering ash (*Fraxinus ornus*) was blown down in March 1907. The Wych elm and a beautiful walnut still flourish, and also the variety of Turkey oak (*Quercus cerris lucumbeana* or *fulhamensis*), so in spite of many disasters Fulham Palace still can show some fine trees.

Chelsea still abounds in gardens. There are the modern plots along the Embankment, laid out with the wriggling path that municipal authorities seem to deem necessary nowadays. The private gardens in front of some of the houses are an older institution, and some can boast of delightful patches of old gardens in their rear also. Behind Lindsay House the Moravian burial-ground is hidden away, and part of its wall may be the actual wall of Sir Thomas More's garden. There are the remains of elms and several good mulberry trees. The large mulberry on the Embankment near looks as if it once might have been in the garden too. Chelsea further possesses one of the first botanical gardens in England, the Duke of York's School with large grass area and fine elm trees, and the spacious grounds that surround the Hospital. Much of the old stately simplicity still clings to these latter, although last century saw many variations in their plan.

The site was occupied by King James's College, founded by Matthew Sutcliffe, Dean of Exeter, in 1610, which, in spite of the King's patronage and the interest of Prince Henry, was a failure. It added to, rather than allayed, religious discussion, and was familiarly known as "Controversy College." The ground was, in 1669, given to the Royal Society, but the buildings were too dilapidated for them to use. To Sir Stephen Fox is probably due the idea of founding a hospital for disabled soldiers, although tradition also attributes some of the credit to Nell Gwynn, who is said to have appealed to Charles II. on their behalf. The King laid the foundation-stone, on the 12th of March 1682, of the building designed by Wren. John Evelyn, as one of the Council of the Royal Society, had been consulted when the idea was first mooted, and in January 1682 he notes in his Diary a talk on the subject with Sir Stephen Fox, who asked for Evelyn's assistance with regard to the staff and management. So in Sir Stephen's study, as Evelyn writes, "We arranged the governor, chaplain, steward, housekeeper, chirurgeon, cook, butler, gardener, porter, and other officers, with their several salaries and entertainments." This list of officials shows the importance of the Garden from the first—and no wonder, as the grounds occupied some twenty-six acres. A survey made in 1702 shows how this space was divided. The largest part, lying to the north of the Hospital, is what is now known as "Burton's Court," and is used as a recreation ground for the soldiers in the barracks near, and a cricket ground for the brigade of Guards. The avenue down the central walk, "planted with limes and chestnuts," was included in the early design, and "Royal Avenue" is a continuation of it, Queen Anne having, it is said, intended to carry it on to Kensington. This part, called "the great court north of the buildings," occupied over thirteen acres. The rest was divided into grass plots between the quadrangle courts and canals, nearly three acres; the "garden on the east, now the gover-

STATUE OF CHARLES II., CHELSEA HOSPITAL

nor's," about two acres; a kitchen-garden towards the
river of more than three acres, two L-shaped canals
with wide walks between, an " apothecary's garden "
for medicinal herbs, bleaching yards, and the churchyard.   The front garden, with its canals in Dutch
style, ended in a terrace along the river.   This garden
was subject to much abuse by the landscape school
of designers.   " It was laid out," wrote one in 1805,
" when the art of landscape gardening was at its
lowest pitch; the principal absurdity in the garden
is cutting two insignificant canals as ornaments, whilst
one side of the garden is bounded by the noble stream
of the Thames."   The writer adds that the gardens
were open on Sundays in summer, and were much
frequented as a public promenade.   These severely-
criticised canals were filled up in the middle of last
century, and the space is now grass with avenues on
either side, and a central obelisk, a monument to our
soldiers who fell in the battle of Chillianwallah.

The statue of Charles II. as a Roman emperor, by
Gibbons, in the centre of the court, was given by Tobias
Rustat.   The view over the simple, spacious garden
from this central court, to the long balustrade with
steps down to the lower terrace, is very satisfying, and
in keeping with the stately architecture.   The Governor's
house has its own special garden, a fine, wide terrace
and large, straight beds, and a delightful red-brick wall
covered with trailing plants and fine iron gateway.   The
old pensioners, in their long coats and weather-beaten
faces, enjoying their " peace pipe " and their well-
earned repose, add very greatly to the picturesque
effect of the Garden, and all its surroundings.   The
churchyard, clearly seen through the railings along

Queen's Road from Chelsea Barracks, has an air of dignified repose. It has been closed since 1854. The first soldier buried there in 1692, Simon Box, had served four kings: Charles I., Charles II., James II.,

GARDEN GATE, CHELSEA HOSPITAL

and William III. The tombs are much worn with age, and it is no longer possible to find some of those known to have been laid to rest there. Among them are two women who had served as privates; one of them, who died in 1739, Christian Davies or "Mother Ross," had served in Marlborough's campaigns. The extraordinary number of centenarians this small burying-ground contains is astounding. William Hisland surely beats the record, as he was married when he was over a hundred! He was born in August 1620, and died in February 1732. Another veteran of 112 died five years later, while another, Robert Comming, who was buried in 1767, was 115, and before the end of the eighteenth century three others, aged respectively 102, 111, and 107, were interred. The eldest of these three, who died in 1772, had fought in the Battle of the Boyne! It certainly speaks well for the care and attention bestowed on them in the Hospital.

The garden to the east of the buildings was part of the original ground, but has had a career and history of its own. It was the famous Ranelagh Gardens, which enchanted the beaux and fair ladies of the eighteenth century. From 1742 to 1803 its glories lasted. Ranelagh House was built by the Earl of that name, who was Paymaster to the Forces in the reign of James II., a clever, unscrupulous person, who amassed considerable wealth in the course of his office-work. He obtained a grant of the land from Chelsea Hospital, built a house and laid out a garden, where the "plots, borders, and walks" were "curiously kept, and elegantly designed." After passing through the hands of his daughter, Lady Catherine Jones, the property was sold to Swift and Timbrell, who leased it to Lacey, the patentee of Drury Lane Theatre. The idea was to turn it into a winter Vauxhall. Eventually it was open from Easter till the end of the summer, and effectually outshone Vauxhall. Walpole, in a letter two days after it was first opened, did not think much of it. " I was there, last night, but did not find the joy of it. Vauxhall is a

little better, for the garden is pleasanter, and one goes
to it by water." Two years later he wrote in a very
different strain. "Every night constantly I go to
Ranelagh, which has totally beat Vauxhall. Nobody
goes anywhere else—everybody goes there. My Lord
Chesterfield is so fond of it, that he says he has
ordered all his letters to be directed thither." Fanny
Burney, in "Evelina," to bring out the character of
the "surly, vulgar, and disagreeable man," makes him
abuse the place which fascinated polite society. "There's
your famous Ranelagh, that you make such a fuss
about; why, what a dull place is that!" The chief
amusement was walking about and looking at each
other, as the poem by Bloomfield puts it—

> "We had seen every soul that was in it,
> Then we went round and saw them again."

The great attraction was the Rotunda, supposed to
be like the Pantheon at Rome. The outside diameter
was 185 feet. An arcade ran all round, and above it a
gallery, with steps up to it through four Doric porticos.
Over the gallery were sixty windows, and the whole was
surmounted by a slate roof. In the middle, supporting
the roof, was a huge fireplace, on the space at first
occupied by the orchestra. "Round the Rotunda,"
inside, were "47 boxes . . . with a table and cloth
spread in each; in these the company" were "regaled,
without any further expense, with tea and coffee." The
whole was adorned with looking-glasses and paintings,
imitation marble, stucco, and gilding. Dr. Arne wrote
music for the special performances; breakfasts were at
one time the rage, and at another masquerades were the
order of the day; while fireworks and illuminations

amused the company at intervals, all through the years
in which Ranelagh was prosperous.

> " There thousands of gay lamps aspir'd
> To the tops of the trees and beyond ;
> And, what was most hugely admired,
> They looked upside-down in a pond.
> The blaze scarce an eagle could bear
> And an owl had most surely been slain ;
> We returned to the circle, and then—
> And then we went round it again."

One of the last entertainments at Ranelagh was the
Installation Ball of the Knights of the Bath in 1803 ;
and a few years afterwards all trace of Ranelagh House,
the Rotunda, and even the Garden was gone. The
ground reverted to Chelsea Hospital, and not a vestige
of the former glories is left. The pleasant shady walks
and undulating lawns on the site, bear no resemblance
to the lines of the former gardens, and only some of the
older trees can have been there when Lord Chesterfield
and Walpole were paying it daily visits.

The most important of Chelsea gardens, and one of
the most interesting in England, is the Physic Garden,
which lies between the Embankment and Queen's Road,
now called Royal Hospital Road. The Garden, both
horticulturally, botanically, and historically, has claims
on every Londoner. England was much behind the
rest of Europe in starting botanic gardens. That of
Padua, begun in 1545, was the first on the Continent,
and it was nearly a hundred years later before any
were attempted in this country. Oxford led the way
in 1632, and the Chelsea one followed in 1673. Its
formation was due to the Apothecaries' Company, and
its first object the study of medicinal herbs. In those

days botany and medicine were closely entwined. Every botanical and horticultural work was occupied with the virtues and properties of plants, far more than their structural peculiarities, or their beauties of form or growth. Gerard, Johnson, and less well-known botanists, were herbalists and apothecaries, so it was only natural that the Worshipful Company of Apothecaries should be the founders of a garden. It was not the first of its kind in London, but it ranks now as the second oldest in England, as its predecessors in London, such as Gerard's Garden in Holborn, and the Tradescants in Lambeth, have long since passed away. It probably, moreover, embodies the earlier one at Westminster, which was under the care of Hugh Morgan, said by his contemporaries to be a very skilful botanist. The Westminster Garden seems to have been still flourishing when the Apothecaries started theirs in Chelsea, but three years later it was bought by them, one of the conditions of sale being that the plants might be moved to Chelsea. The land in Chelsea was leased from Lord Cheyne. By the time the lease had expired, Sir Hans Sloane was owner of the property, having purchased it from Lord Cheyne in 1712. He granted the land to the Apothecaries' Company on a yearly rent of £5, on condition that it should always be maintained as a Physic Garden, and certain other conditions, such as supplying a number of specimens to the Royal Society. The deed of gift further provided that should the Apothecaries not continue to fulfil their obligation, the Garden should be held in trust by the Royal Society, and should they not wish to take it over, by the College of Physicians. It was acting in conformity with these wishes, that, when the Apothecaries ceased to desire to maintain it, the Charity

Commissioners, in 1898, established a scheme for the management of the Garden: £800 towards its maintenance was provided by the London Parochial Charities, who became trustees of the Garden, and £150 by the Treasury. A committee was appointed to manage the Garden, and see that it fulfilled the founder's intentions. The original societies mentioned by Sir Hans Sloane, the Treasury, the London County Council, and other modern bodies each nominate one representative on the board of management, and the trustees appoint nine. It has been worked under this scheme since May 1899. The buildings and green-houses, which were tumbling down, have been rebuilt, and now include up-to-date conveniences for growing and rearing plants, and a well-fitted laboratory and lecture room. The Garden is certainly now fulfilling the purposes for which it was founded. It has proved to be of the greatest use to the students of the Royal College of Science, and members of schools and polytechnics. Cut specimens, for demonstration at lectures, are sent out in quantities during the summer, often as many as 750 in a day. Students and teachers have admission to the Garden, and the numbers who come (nearly 3000 is the average annual attendance) show it is appreciated. Lectures on advanced botany have been attended by an average of seventy students, and research experiments are carried on in the laboratory. Seeds are exchanged with botanical gardens all over the world, to the extent of over a thousand packets in a year. In this it is carrying on a very early tradition, as seeds were exchanged with the University of Leyden in 1682, after Dr. Herman, from that city, had visited Chelsea.

Even in its early days the Apothecaries found the

Garden expensive to keep up. When in 1685 it cost
them £130, besides the Curator's salary, they made an
arrangement, by which they paid him £100 a year, out
of which he had to keep up the Garden, and was allowed
to sell the plants. Watt was the first Curator under this
new plan, and Doody, a botanist of some standing who
succeeded him, was under the same conditions. Philip
Miller was appointed Curator, after the land had been
given by Sir Hans Sloane, and other well-known men
have been connected with it. After 1724, besides the
Curator, a "Præfectus Horti," or Director, was appointed
to visit and inspect the Garden, and report on its condition to the Company. Sometimes there was a little
rivalry between the two, and at one time this occasioned
two lists of the plants contained in the Garden being
published, one by Isaac Rand, the other by Philip Miller.
Among the famous names in botany or horticulture connected with the Garden are Dr. Dale, Mrs. Elizabeth
Blackwell, James Sherard and his brother William,
Joseph Millar, William Curtis, Forsyth, Robert Fortune
and Dr. Lindley, and Nathaniel Ward, the inventor of
"Wardian Cases." But of all the Curators, Philip
Miller was one of the most eminent, and did most
for the Garden. His Dictionary was for years the
standard work on horticulture, and went through
numerous editions and translations. He published a
catalogue of plants in the Physic Garden in 1730.
The last "Præfectus Horti" was Lindley, who held
the office from 1835 to 1853. During that time the
expenses were getting too heavy for the Society, and
after his death no successor was appointed. Thomas
Moore, who was co-editor with Lindley of the well-
known "Treasury of Botany," and author of several

works on British ferns, continued alone as Curator. He held the office from 1848 to 1887. During his later years the Garden gradually declined for want of funds, and after his death no new appointment was made by the Apothecaries, and a labourer looked after the grounds. With the advent of the new authority and great expansion of work, the office was once more bestowed on a competent man, William Hales, the present Curator, who ably maintains the old traditions of the garden.

One of the institutions of early days which has had to be discontinued was the "herborising." Expeditions in search of herbs were undertaken by the students, in company of their teacher, in the neighbourhood. After 1834, owing to the spread of London, these excursions had to be abandoned.

The famous cedars were planted in Watt's time, and from contemporary references to them, there seems no doubt that they were the first to be grown in England. John Evelyn in his "Sylva" in 1663, writing of the cedar, says, "Why should it not thrive in Old England?" and Ray is astonished in 1684 to see the young trees flourishing at Chelsea without protection. They are shown in a plan of the Garden in 1753 (the year of Sir Hans Sloane's death) at the four corners of a pond, which no longer remains in the same position. Eighteen years later the two furthest from the river were cut down (1771), "being in a decayed state" (and no wonder) from the rough usage they had been subjected to. The timber, 133¾ feet, was sold at 2s. 8d. a foot, and, together with the branches, the trees fetched £23, 9s. 8d. The two specimens nearest the river were for nearly a hundred years a conspicuous object, although much injured by snow in 1809. By 1871, only one remained, and, in a

x

report of the Garden seven years later, it was said to be in a "dying condition." At the time the new Management Committee came into office, that one was quite dead. They left the tree standing until the fungi on it bècame a danger to the rest of the trees in the Garden, when most reluctantly it was felled in March 1904, all the sound parts of the timber being carefully preserved. Miller gives a good account of them in his time. "The four trees," he writes, "(which as I have been credibly informed) were planted there in the year 1683, and at that time were not above three feet high; two of which Trees are at this time (viz. 1757) upwards of eleven feet and a Half in girt, at two Feet above ground, and thereby afford a goodly shade in the hotest Season of the Year." He goes on to point out that they were planted so near the pond, which was bricked up to within two feet of them, that the roots could not spread on one side. Whether the water was good for them he is not sure, but feels certain it was injurious to cramp the roots. The two specimens nearest the green-house had had some of their branches lopped off, to prevent their shading the grass, and suffered in consequence. Though one remained for nearly 150 years after Miller gave these measurements, it was only 13 feet round the trunk at the base when it was felled, and was so completely rotten it must soon have fallen. Miller records that three of the trees began producing cones about 1732, and that in his time the seeds ripened, and germinated freely, so it is probable that many plants in England are descendants of the Chelsea trees. That these were actually the first to be grown in England there is not much doubt. Evelyn regrets in his "Sylva" the absence of the cedars in England. The only trees which have put forth rival claims to the Chelsea ones are those of Bretby and Enfield. The Bretby one is undoubtedly very old, but there is no early reference to it in histories which mention the Enfield trees, and the famous one at Hendon, traditionally planted by Queen Elizabeth and blown down in 1779, and a few others ; and there is no contemporary evidence of the date of its planting to warrant the assumption that it was before 1683. The Enfield tree in the garden of Robert Uvedale was said, in 1823, by Henry Phillips, to be about 156 years old, therefore older than the Chelsea ones by some six years; but there is no evidence to corroborate this. When Gibson describes the Garden in 1691, he makes no mention of it, and it seems unlikely he would have omitted such an important tree. There exists much correspondence with Uvedale and botanists of his time, but in none of the letters or early notices is the cedar mentioned before Ray's note of the Chelsea trees, or even referred to as the first planted in England, so it seems the Chelsea trees' claim to be the first is fairly established.

The oriental plane, which fell just as it was going to be taken down in 1904, was one of the finest in London, planted by Philip Miller, and is quoted by Loudon, in 1837, as then 115 feet high. Some of the other famous trees have also died, such as the cork trees and paper mulberries ; but some have been more fortunate, and are among the oldest of their kind in England. The *Koelreuteria paniculata* is probably the finest in this country, and the other old trees which were noted as being particularly fine specimens in 1813 or 1820, and which are still alive, are *Diospyros Virginiana*, the Persimmon or Virginian date plum, the Quercus ilex, black walnut, mulberry, and *Styrax officinale*. *Rhus juglandifolia*, which

grows by the wall, was probably planted when introduced from Nepaul in 1823. The wistaria and pome-
granate are old and still flourishing, and young plants
of the trees once famous in the Garden are doing
well. The amount of attention the novelties in the
Physic Garden used to attract is well shown by the
spurious translation of De Sorbière's travels. The little
book, published in 1698, purported to be a translation
of De Sorbière, but was really an original skit. The
writer pretends De Sorbière visited the Garden, and re-
ported a delightful series of imaginary flowers. "I was
at Chelsey, where I took particular notice of the plants
in the Green House at that time, as *Urtica male oleus
Japoniæ*, the stinking nettle of Japan ; *Goosberia sterelis
Armenia*, the Armenian gooseberry bush that bears no
fruit (this had been potted thirty years) ; *Brambelia fruc-
tificans Laplandiæ*, or the Blooming Bramble of Lapland ;
with a hundred other curious plants, and a particular
Collection of Briars and Thorns, which were some part
of the curse of the Creation." That it was worth while
laughing at the Garden in a popular skit, shows what
an important position it had taken. The green-houses
were among the earliest attempted, and many scientific
visitors describe their plans and arrangements. They
were rebuilt at great cost in 1732. The statue to Sir
Hans Sloane, by Michael Rysbrach, stood in a niche in
the green-house wall. It was moved to the centre of the
Garden in 1751, where it still stands. The Garden was
honoured by a visit from the great Linnæus in 1736, and
he noted in his diary : "Miller of Chelsea permitted me
to collect many plants in the Garden, and gave me several
dried specimens collected in South America." Among
the valuable bequests to the Garden were collections of

CHELSEA PHYSIC GARDEN

dried plants, now in the British Museum of Natural
History, and a library left by Dr. Dale in 1739, on condition that "suitable and proper conveniences" were
made for them at the Physic Garden. They should be
there still, and the new buildings are eminently suited
for their reception ; and their use to students would be
very great, now that the Garden is well equipped for
supplying all the requirements for the modern teaching
of botany.

Before quitting these gardens of historic interest,
there is one which must not be forgotten, although its
former charms have vanished, and it can no longer claim
such botanical curiosities as the Chelsea Physic Garden—
that is, the remains of John Evelyn's Garden of Sayes
Court. The Garden is now enjoyed by numbers in that
crowded district of Deptford, through the kindness of
Mr. Evelyn, the descendant of the famous diarist, John
Evelyn, who keeps it up as well as opens it to the public.
The Manor of Deptford was retained by the Crown in
James I.'s time, and Sayes Court was leased to Christopher Browne, the grandfather of Sir Richard Browne,
whose only daughter and heiress John Evelyn married.
After his wife had succeeded to the property, and they
had lived there some years and made the Garden, John
Evelyn purchased the freehold land from Charles II.
The delight he took in his garden, how he exchanged
seeds and plants, imported rare specimens from abroad,
through his many friends, and grew them with success,
is well known. The ruthless way his treasures were
treated by Peter the Great was a sore trial to Evelyn.
The Czar amused himself, among other acts of van-
dalism, by being wheeled about the beds and hedges
in a wheelbarrow. The holly hedge, even, he partially

destroyed. In writing of the merits of holly in his "Sylva," Evelyn says of this one : " Is there under heaven a more glorious and refreshing object of the kind, than an impregnable Hedge a hundred and sixty feet in length, and seven feet high, and five in diameter, which I can shew in my poor Gardens at any time of the year, glittering with its armed and vernish'd leaves ? the taller Standards at orderly distances blushing with their natural Corall. It mocks at the rudest assaults of the Weather, Beasts, and Hedgebreakers." This hedge has long since departed, but young hollies, planted in groups on the same part of the Garden, keep up the old associations. One wing of the house is standing, and is at present used as a school. The walled garden on the south side is still there, and on the north a wide terrace walk, with a straight grass lawn with large beds, is in keeping with the old place. But instead of the views over the river, and the Garden descending to the water's edge, there is a high rampart of the buildings of the Foreign Cattle Market, from whence the sounds of lowing oxen mingle with the din of streets which close round the Garden on the three other sides. In spite of these drawbacks, it is delightful to know, that the surviving portion of the once-beautiful Garden is fulfilling a want among the poor in a way that would have appealed to the generous and kind-hearted author.

These are some of the chief gardens of historic interest, but it by no means exhausts the list of the smaller ones rich in associations, green courts attached to schools, almshouses, hospitals, or such-like, which are hidden away in unexpected corners throughout London.

CHAPTER XIII

PRIVATE GARDENS

Even in the stifling bosom of a town
A garden, in which nothing thrives, has charms
That soothe the rich possessor ; . . .

—Cowper.

IN writing of the private gardens of London it is difficult to know where to begin. There are a few large and beautiful gardens, but for the most part the smaller they are, and the less there is to write about them of interest to the general reader, the more they are of value to the happy possessors. It is the minute back-garden, invaded by all the cats of the neighbour-hood, with a few plants on which an infinity of time and trouble, care and thought, have been expended, that is the real typical London garden. What a joy to see the patches of seeds come up in the summer, and with what expectation are the buds on the one lilac bush examined to see if really at last it is going to flower ! What pleasure the fern dug up on a summer holiday gives, as it bravely uncurls its fronds year by year ! What delight is occasioned if the Virginian creeper, which covers the wall, grows more luxuriantly than those of the houses on either side, and what excite-

327

328 LONDON PARKS & GARDENS

ment if it really turns red once in a way in October, instead of shrivelling up to an inglorious end! What grief is felt when the fuchsia, purchased as a fitting centre-piece to the formal geranium bed, loses its buds one by one before they expand! These and many similar joys and sorrows are the portion of those who tend small gardens in London. How fascinating it is to look into back-gardens as the train passes over viaducts out of the heart of the town. Certainly the differences in their appearance show what skill and devotion can accomplish. Nothing but real love of the plants, and a tender solicitude for their welfare, can induce them to exist in the confined areas and stifling atmosphere of the average London garden. But even these inauspicious surroundings may be brightened by flowers. When those absolutely ignorant of the requirements of plant life take to gardening in the country, they have Nature at hand to help them. The sunlight, air, and good soil supplement their deficiencies of knowledge, and, though terribly handicapped by careless planting, unsuitable situations, want of water, and such drawbacks, the plants can struggle with success to maintain their natural beauty. But let the ignorant try in town to grow plants, where all the conditions militate against them, instead of assisting, and the results are very different. For instance, many a small back-garden, or even window box, is planted year after year with no renewal of the soil. The crumbling mould, which is either caked hard or pours like dust from the hand, is completely exhausted, and the poor plants are starved. They should be given plenty of what in gardeners' slang is called " good stuff," if they are to grow in such adverse conditions. A little of the money expended on plants which dwindle and die, spent on manure or good soil, would better repay the would-be gardener. Many plants require a good deal of water when making their growth, and if that is denied them they will not thrive, no matter how great the solicitude for their welfare in other ways. Washing the leaves, especially of evergreens, and scrubbing stems is also a great help, as leaves choked with dirt have no chance of imbibing the life-giving properties necessary to the plant.

The back-garden has many enemies besides soot and fogs. Cats are one of the greatest trials, and most destructive. Sparrows also are very mischievous. They will pick the flower-buds off trees just at the critical moment. A wistaria climber laden with young blossoms they will destroy in a few days, just before the purple buds appear. But, notwithstanding all these pests and difficulties, it is surprising how many things will not only survive, but grow well. The task becomes more and more easy as the houses recede from the City. In St. John's Wood, Bayswater, or Earl's Court, in Camberwell or Stoke Newington, plants will grow better than in Bloomsbury or Southwark. But yet it is possible to grow many things even in Whitechapel.

It is impossible to prescribe the best plants for all London gardens, as there is such a great difference in soil and aspect, that what does well in one part will not flourish in another. The heavy soil of Regent's Park, for instance, is well suited to peonies, which do not seem at home in Chelsea. On the other hand, some of the showy, hardy spring flowers, such as wallflowers and forget-me-nots, die off with fogs much more quickly in the Regent's Park than in other districts. Any deciduous tree or shrub thrives better than an evergreen

or a conifer in any part of London. The fresh growth
of clean leaves every year, by which the plant absorbs
much of its nourishment, must necessarily be better
for it than dried-up, blackened leaves. Among flowering shrubs, a great number grow sturdily in London.
Laburnums of all kinds, thorns in many varieties, flower
well; lilacs grow and look fresh and green everywhere,
but cannot be depended on always to flower; almonds,
snowy medlars, double cherries, weigelas or diervillas
succeed; broom, Forsythias, acacia, syringa, many kinds
of prunus, ribes, rose acacia, Guelder rose, Japanese red
peach, *Kerria japonica*, *Hibiscus Syriacus*, or *Althæa frutex*,
are all satisfactory, and many more could be mentioned.
Yucca gloriosa will stand any amount of smoke, and
Aralia spinosa does well in many parts; and among
evergreens, *Arbutus Andrachne* can be recommended.
Fruit-trees, pears, and apples are charming when in
bloom, and in a large space, or to cover a wall, figs
are valuable.

Alpines grow astonishingly well, and though a considerable percentage will die from the alternating damp
fogs and frost in the winter, many will really establish
themselves, and be quite at home, much nearer the
heart of London than Dulwich, where many have
been cultivated. "I know a bank whereon the wild
thyme grows" in London—not a green, mossy bank,
but rather a blackened rockery; still the slope is
really covered with large patches of wild thyme, purple
with bloom in the summer, carefully marked by the
London County Council "*Thymus serpyllum*," for the
benefit of the inquiring. Several of the other thymes,
which form good carpets, will also grow. *Antennaria
dioica*, a British plant, forms a pretty silvery ground-

work on beds or rockeries, and nothing seems to kill it. Saxifrages in great numbers are suitable, beginning with the well-known mossy green *hypnoides*, to the giant known as *Megasia cordifolia*, also sedums, sempervivums, aubrietias, phloxes, tiarella, dianthus in variety; and several other Alpines have succeeded in different parks and gardens, such as *Androsace sarmentosa*, *Dryas octopetala*, yellow fumitory, *Cotoneaster frigida*, the small ivy *Hedera conglomerata*, *Achillea tormentosa*, *Lychnis Haageana*, *Linnæa borealis*, *Azalea procumbens*, *Campanula garganica*, only to mention some that have been noticed; even edelweiss has been successfully grown in the centre of London.

A few annuals will make a good show, and nothing is better in a window-box or really dingy corner than Virginian stock; but, as a rule, it repays trouble best to rear perennials. Seedling wallflowers, sweet Williams and Canterbury bells, and such like, make a border bright. The great secret of success in growing annuals is to thin them out well; the patches of seedlings are too often left far too much overcrowded. This "thinning" is even more important than good soil and careful watering. Marigolds thrive best of all, and will often seed themselves, but a few other annuals can be safely recommended.

Candytuft.
Catchfly (Silene pendula and armeria).
Erysinum perofskianum (a kind of Treacle mustard).
Eschscholzia.
Flax (scarlet).
Godetias.
Ionopsidium acaule (violet cress).
Larkspur (annual).
Love-in-a-mist (Nigella).
Nasturtiums.
Phlox drummondi.
Snapdragon (Antirrhinum).
Toadflax (Linaria).

Very many things may succeed well that are not specially noted here, but the following list of fifty herbaceous plants have all been seen really growing, and coming up, year after year, in private gardens in London. Some are not so sturdy as others; for instance, neither alyssum nor phlox flourish as well as thrift or the members of the iris tribe, but all are hardy in London. Thomas Fairchild, who had a famous nursery garden at Hoxton, writing of City gardens in 1722, gives his experience of plants that succeed best, and many on his list are those that do well still. He specially notes some growing in the most shut-in parts of the City, which were flourishing: fraxinella in Aldermanbury, monkshood and lily of the valley near the Guildhall, bladder senna in Crutched Friars, and so on, mentioning many of those which still prove the most smoke-resisting. One large, coarse, but handsome plant deserves mention, as it grows so well it will seed itself, and that is the giant heracleum. It propagates itself in the garden of Lowther Lodge, Kensington Gore, and in much more confined spaces, even in the garden used by the London Hospital, near the Mile End Road.

LIST OF FIFTY HERBACEOUS PLANTS

Alyssum.	Comfrey.
Auricula.	Crane's bill.
Bachelors' buttons.	Creeping Jenny.
Buglos.	Crown Imperial.
Campanula—several varieties.	Cyclamen.
Candytuft.	Day lilies.
Carnations.	Dictamnus fraxinella (burning bush).
Centaurea.	Doronicum (leopard's bane).
Chrysanthemums.	Erigeron (Fleabane).
Columbines.	Funkias (Plantain lilies).

Galega officinalis.
Golden rod (solidago).
Heucheras.
Hollyhocks.
Iris—several varieties, especially those with rhizomes and non-bulbous roots.
Japanese anemone.
Larkspur.
Lilies of the valley.
Lilies—
  Canadense.
  Candidum.
  Davuricum.
  Lancifolium (speciosum).
  Martagon dalmaticum.
  Pyrenaicum.
  Tigrinum.
London Pride (also many other Saxifrages).
Lupin.
Mallow.

Michaelmas daisies.
Monkshood.
Montbretia.
Pansies.
Periwinkle.
Phlox.
Polygonum.
Primroses (also Japanese primulas, cowslips, and polyanthus).
Pyrethrum.
Rock roses.
Solomon's seal.
Southernwood.
Speedwell (Veronica amethystina and others).
Spiræa (S. aruncus, venusta, &c.).
Sunflower (perennial, including Harpalium).
Thrift.
Tradescantia.
Trollius.

Of climbing plants the Virginian creeper, which makes a green bower of so many London houses, must come first, but the real grape vine is quite as successful. In several parts of London vines laden with grapes may be seen in the autumn, by those on the look-out for such things. One vine in Buckingham Gate had forty bunches of fruit that ripened in 1906. On one branch of a vine, near Ladbroke Square, fourteen purple bunches were hanging in a row at the same time, and in other parts of the town well-cared-for vines will bear well. Wistaria also thrives, and jasmine, yellow or white, and ivy. Besides these in constant use, for more special gardens there are Everlasting peas, Dutchman's pipe (*Aristolochia*), clematis, Jackmani, Montana, or the Wild

Traveller's Joy, and Passion flower; also convolvulus, *Cobæa scandens*, and gourds of all kinds for the summer.

Spring flowers planted in autumn succeed, and even those in pots or boxes in windows or on roof gardens flower freely. Hyacinths, crocus, tulips, daffodils, and narcissus do well; snowdrops are not so successful as a rule, but Spanish Iris will make a good show when the earlier bulbs are over. The minute green-house which often opens out of a staircase window in London houses can easily be made gay in spring by this means. Acorns and chestnuts sown in the autumn in shallow pans and covered with moss make a delightful small forest from May onwards. Foxgloves dug out of the woods will flower well in these dingy little green-houses, and are a delightful contrast to the ferns which will flourish best in them. A few other plants are sturdy for this purpose, such as the fan palms, *Chamærops excelsa*, *Fortunei*, and *humilis*, Aspidistra, *Aralia Sieboldii*, *Selaginella Kraussina*, the Cornish money-wort (Sibthorpia). Geraniums will flower well, and Imantophyllums (or Clivias) are one of the most accommodating plants for such small green-houses, as although they take up an undue share of room on account of the large pots necessary, they will flower well every year.

Roses only do fairly well; but though they sometimes will last two or three years, they are apt to give disappointments and must often be renewed. The climbing roses, however, in some gardens are very charming. In one of the prettiest in London—that belonging to Sir Laurence Alma-Tadema, in Grove End Road—the illustration shows how charmingly an iron trellis is covered with red and white roses. The garden

THE GARDEN OF SIR LAURENCE ALMA TADEMA, R.A.

is most artistically arranged and is a good illustration of how much can be made of a small space. A large evergreen oak overhangs the basin with a stone margin and splashing fountain, on which water-lilies gracefully float. The variety and harmony of the whole garden, with its paths shaded by fig-trees, apples and pears, cherries and lilacs, sunny borders with Scotch roses, Day lilies, foxgloves, and iris, and formal fountains, all in a small space, yet not crowded, and bright with flowers, is delightful. Another small garden in Kensington—tended by Lady Bergne—of quite another type, contains nearly all the flowers that have been mentioned as growing well in London. It is only the stereotyped long narrow strip at the back of the house; but by putting a path and rock-work and pools of water on one side, and having grass and flower borders on the other, backed by flowering shrubs and ferns at the shaded end, a great variety of plants have been grown successfully.

In most London gardens very little enterprise is shown. The old system of bedding out is adhered to. Of the large London houses standing by their own lawns, none have gardens of any horticultural interest. Montagu House is on the site of the extensive lgardens of Whitehall, and the present lawn is where the bowling green, with its gay throng of players, lay in former years, and the terrace keeps up the tradition of the wide terraces that descended from the palace to the green. The turf is still fair and green, and is brightened in summer by lines of geraniums, white daisies, and calceolarias. Devonshire House garden, on the site of the famous one belonging to Berkeley House that covered all the present Square, is in the same way merely planted with the usual summer bedding plants.

Lord Portman's house, 22 Portman Square, is where
Mrs. Montagu, the Queen of the Blue-Stockings, held
her court.   The present garden, with spacious lawn,
has no horticultural peculiarity, but its historical interest
lies in the fact that it was here that Mrs. Montagu
entertained the chimney-sweeps, every year on the 1st of
May.   She is said to have done so, to give these poor
children "one happy day in the year," and when the
horrors and tragedies attending the lives and often
deaths of these cruelly treated little creatures is realised,
it is not to be wondered at that one lady was humane
enough to befriend them.

A quaint pathetic poem by Allan Cunningham,
written in 1824, records in characteristically stilted
language an incident supposed to have occurred to
Mrs. Montagu.   A sad boy, whose life was spent in
climbing flues, is pictured, and one lady he supplicates
turns away—"And lo! another lady came," and spoke
kindly to him, asked him why he thus spent his life,
listened to his tale of how he was an orphan and "sold
to this cruel trade."

> "She stroked the sooty locks and smiled,
>   While o'er the dusky boy,
> As streams the sunbeam through a cloud,
>   There came a flash of joy.
> She took him from his cruel trade,
>   And soon the milk-white hue
> Came to his neck ; he with the muse
>   Sings, ' Bless the Montagu.' "

Her kindness is recorded in other poems, and in
her lifetime took the practical shape of a sumptuous
spread of beef and plum-pudding on the lawn of her
house in Portman Square.

Grosvenor House garden, with terrace and lawn sloping down to large trees, has natural advantages for a beautiful garden, but a row of beds along the terrace are the only flowers. The owners of these large London gardens have such an abundance of floral display elsewhere that no real gardening seems to be attempted. To understand what are the horticultural possibilities of London, it is in the minute back-garden that the lesson must be learned, and the subject studied. Holland House is an exception to this rule, for there the most beautiful garden, in keeping with the magnificent old house, is kept up, and the greatest care and skill were bestowed on it with wonderful results by the late Earl of Ilchester.

No house, perhaps, has more associations than Holland House. Its history has been so often written, that to go over it in detail would be superfluous. Built by Sir Walter Cope, while Elizabeth was on the throne, from the designs of Thorpe, it doubtless from the first had a good garden, as in those days great care was expended on the surroundings of a house, for people realised, as did Bacon, that, "men come to build stately, sooner than to garden finely; as if gardening were the greater perfection." The second stage in its history, when it passed to Henry Rich, through his marriage with Sir Walter Cope's daughter and heiress, was even more eventful. He enlarged the house, which became known as Holland House after Charles I. had created him Baron Kensington and Earl of Holland. His wonderful personal charm, inherited from his mother, the "Stella" of Sir Philip Sidney, made him a general favourite; but not even his attachment to the Queen preserved him from disloyalty, although in the end he

Y

fought for the King's cause. While he was on the Parliamentary side, Holland House was often the meeting-place of its leaders. Cromwell and Ireton talked together in the centre of the field in front of the house, so that their raised voices, occasioned by Ireton's deafness, should not be overheard. For a time after the Restoration, Holland House was tenanted by various people of note, to whom it was let out in suites by the widowed Countess. One among them, the Frenchman Chardin, who became famous by his travels to Persia, it has been surmised, may have brought some of the rare plants to the garden. The connection with Addison came from his marriage with the Dowager Lady Warwick, to whom the house belonged, the second Lord Holland having succeeded his cousin as Earl of Warwick. He must have delighted in the gardens of Holland House, although they were hardly so wild as the ideal one he describes in the *Spectator*. There he said, "I look upon the pleasure which we take in a garden as one of the most innocent delights in human life." No doubt he found some solace in the beauties of Holland House garden to cheer the depression of the unhappiness the marriage had brought him. The brilliant days of Holland House continued after it changed hands, and was owned by Henry Fox, second son of Sir Stephen Fox, who was chiefly instrumental in starting Chelsea Hospital. Henry Fox eloped with Lady Caroline Lennox, and was afterwards created Lord Holland. He took great interest in his garden, and was advised and helped by the well-known collector and horticulturist, Peter Collinson. This friend was the means of introducing many new plants to this country—a

genus Collinsonia was named after him—and he must
have been pleasant and good besides, for his biographer
says to him was attached "all that respect which is due
to benevolence and virtue." He was in correspondence
with leading men in America, and was constantly receiving seeds and plants, and his own garden contained "a
more complete assortment of the *orchis* genus than,
perhaps, had ever been seen in one collection before."
No doubt some found their way to the gardens of his
friend, Lord Holland. How astonished they both would
be could they peep for a moment at the orchids displayed in the tents of the Horticultural Society's shows,
which have been allowed to take place in the park
where Cromwell conversed? At this time the gardens
must have been considerably remodelled, as the taste
for the formal was waning, and the "natural" school
taking its place. One of the pioneers of the natural
style, Charles Hamilton, assisted the new design. His
own place, Painshill, near Cobham, in Surrey, embraced
all the newest ideas, groves, thickets, lakes, temples,
grottos, sham ruins, and hermitages. A contemporary
admirer, Wheatley, says of Painshill, it "is all a new
creation; and a boldness of design, and a happiness of
execution attend the wonderful efforts which art has
there made to rival nature." No doubt this adept in
the new art would introduce many changes. The
"Green Lane" was a road shut up by Lady Holland,
and Hamilton is said to have suggested turfing it. He
appears to have been fond of woodland glades and turfed
the shaded walks in his own creation, so it seems very
likely that the idea of grass was his. In the Green Lane,
Charles James Fox, son of the first Lady Holland, who
closed the road, loved to walk, and still the Green Lane

is one of the most attractive spots in all London. The
fame of Holland House increased as time went on, and
some of its most brilliant days were during the time
of the third Lord Holland, when Lady Holland drew
all the wit and fashion of London to her salon.
Although it is no longer a country place, and though
no highwaymen have to be braved to reach it, and
though its surroundings are completely changed, the
garden of Holland House was never more beautiful
than it is to-day. It is easy to forget it is a London
garden, the flowers look so clean and fresh. The long
vista into the rose garden from the lawn, which lies
to the north, is flanked on either side with pink roses,
that pretty free-flowering Caroline Testout. To the
west, overlooking the Dutch garden, the view is even
more attractive, and the garden so well harmonises with
the house that it is easy to picture the beaux in wigs,
and ladies in hoops and powder, moving among the
box-edged beds. On the south, the wide terrace shown
in the sketch was made in 1848, when the footpath was
altered and the entrance to the house changed to the
eastern side. The stone basin in the centre was put
in by the late Lord Ilchester. The hybrid water-lilies,
raised by Marliac, grow well in it, and that rather
delicate, but most beautiful of the Sagittarias, *monte-
vidensis* has flowered there. The raised terrace on the
arches of the old stables, which encloses one side of
the garden and is covered with a tangle of ivy, affords
a charming view over the Dutch garden. Beyond
is the old ballroom, orangery and garden enclosed by
arches of cut limes. A terrace runs to the south of
the Dutch garden and orangery, and the Italian garden
which lies here is in itself as complete a contrast to the

THE LILY POND, HOLLAND HOUSE

box-edged beds of the Dutch garden as is the Japanese
garden, a new addition which lies further to the north.
It was near here that the fatal duel between Lord
Camelford and Colonel Best took place in 1804. There
is yet another small enclosed garden cut off by thick
yew hedges and fat hollies from the rest. In it is the
seat inscribed with lines to the poet Rogers :—

> "Here Rogers sat, and here forever dwell
> With me those Pleasures that he sings so well."

In this garden, year by year, dahlias have grown
ever since they were first successfully grown in England.
In 1789 the dahlia came for the first time from the
New World to the Old. It was then sent to Spain,
and that same year Lady Bute procured some from
Madrid. She was not, however, successful in growing
it and it quite died out, until it was reintroduced by
Lady Holland in 1804. The plants remained rare in
England for some years. It was being grown in France,
Germany, and Holland, but little had been done to
improve the original plant. When, however, a larger
supply was available in England after 1814, the English
growers took it up, and produced, before long, the
round very double flowers which soon became the rage.
In stilted style a writer in 1824 describes the dahlia
mania, after giving the history of its introduction. "It
was left to English capital and perseverance," he says,
"to illuminate the northern part of the globe by the
full brilliancy of these floral luminaries." Thus in
extravagant language he continues to sing the praises
of the dahlia. It is curious that the name is now
generally pronounced as if it were "dalea," forgetful
of the fact that there is a flower, something like a vetch,

called "Dalea" by Linnæus, after Dr. Samuel Dale, who died in 1739, a well-known botanist and friend of Ray. The dahlia was named long after in honour of the Swedish botanist Dahl.

The so-called "Japanese garden" was made by the late Lord Ilchester. It is extremely pretty, but is entirely an English idea of what a Japanese garden is like, and, however pleasing it may be to the uninitiated, would probably shock the Japanese gardener, who is guided by as precise rules in his garden, as the painter in his art. In Japan the rules governing the laying-out of a garden are so exact that, apparently, it requires years of study to acquire the rudiments. The Japanese garden at Holland House, which is pleasing to the English eye, consists of a little stream descending through grassy lawns, with groups of plants, a stone lantern, and rustic bridges, and water plants at each little pond. The delightful *Iris kæmpferi* flowers well, and yuccas, which, by the way, come from America, and not Japan; neither do *Aralia spinosa* or *Saxifraga peltata*, which together form charming groups, with auratum lilies in the summer and other Japanese plants. The French hybrid water-lilies, of varying shades of pink, red, and yellow, here too make a picture, with their brilliant blossoms floating on the miniature pools —while bamboos, maples, and eulalias, true natives of Japan, make a soft and feathery background. Above the Japanese garden there is a well-furnished rock garden, and between that and the roses, which make such a grand display on the north of the house, green walks through rhododendrons and flowering shrubs unite the gardens. There are some really fine trees, as well as all the charming flowers, in the grounds.

Near the bridge leading to the Japanese garden there is a beautiful evergreen oak and rare forest trees, while on the lawn some old cedars, planted by Charles James Fox, are showing signs of decrepitude, although the delightful picturesque effect a cedar always has, adds one more to the many charms of this, the most beautiful as well as the largest of London gardens.

There is a charming group of houses standing in their own grounds still left on Campden Hill, although Campden House has been demolished and its site built over within the last few years. The property on which Campden House stood, and some authorities say the house itself, was won over some game of chance in James I.'s time by Sir Baptist Hicks, afterwards Viscount Campden, from Sir Walter Cope, the builder of Holland House, hard by. It was to Campden House that Queen Anne's little son, the Duke of Gloucester, was taken for country air. The air is still pleasant on these heights, and the open tract of Holland Park gives so much freshness that plants flourish wonderfully. There are good gardens attached to many of the houses—Cam House, Blundell House, Aubrey House, Thornwood, Holly, and Moray Lodges, and several others. Holly Lodge is noteworthy as having for a few years been the residence of Lord Macaulay. There are some charming trees in the grounds, even yews (which are among the first to suffer from smoke) looking well; a good old mulberry and silver elms, and a camellia in a border near the wall, which often flowers out of doors, although some years the half-open buds drop off from the effects of frosty fogs.

Cam House has one of the most charming gardens. It is now lived in by Sir Walter Phillimore, and has

been in his family for some 150 years. It was well
known as Argyll Lodge, as the late Duke bought the
lease and made it his town residence from the time
he first took office in Lord Aberdeen's ministry in 1852.
Before that it was known as Bedford Lodge, as the
Duchess of Bedford, step-mother of Lord John Russell,
the Prime Minister, had lived there and laid out and
planted most of the garden. The "two very old oaks,
which," wrote the Duke of Argyll, "would have done
no discredit to any ancient chase in England," are still
to be seen. The Duke was also delighted with the wild
birds which there made their homes in the garden; in
fact, he says in his Memoirs, it was the sight of the
"fine lawn covered with starlings, hunting for grubs
and insects in their very peculiar fashion," the nut-
hatches "moving over the trees, as if they were in
some deep English woodland," the fly-catchers and the
warblers, that made him decide to take the house.
During the half-century he lived there many of the
birds, the fly catchers, reed-wren, black cap, and willow-
wren, and nut-hatches, deserted the garden, but even
now starlings and wood-pigeons abound, and, what is
even more rare in London, squirrels may be seen
swinging from branch to branch of the old trees. Besides the two old pollard oaks there are good beech and
copper-beech, elder, chestnuts, snowy medlar, sycamore,
several varieties of thorn, and a large Scotch laburnum,
Laburnum alpinum, which flowers later than the ordinary
laburnum, and is therefore valuable to prolong the
season of these golden showers. The leaves are broader
and darker, and growth more spreading. On the vine
trellis is a curious old vine with strongly scented flowers.
All the plants which thrive in London are well grown

in the charming formal garden and along the old wall, which is covered with delicious climbing plants. So luxuriously will some flowers grow, that the hollyhocks from this garden took the prize at the horticultural show held in the grounds of Holland House, in a competition open to all the gardens in the Kingdom.

At Fulham there is a charming garden, with trees which would be remarkable anywhere, and appear still more beautiful from their proximity to London. These trees in the grounds of Broom House have fared on the whole better than those at Fulham Palace, hard by. It is separated from the Palace by the grounds now attached to the club of Hurlingham. Of Hurlingham there is not much early history. Faulkner, the authority for this district, writes in 1813: "Hurlingham Field is now the property of the Earl of Ranelagh and the site of his house. It was here that great numbers of people were buried during the Plague." The same authority mentions: "The Dowager Countess of Lonsdale has an elegant house and gardens here in full view of the Thames," and Broom House is shown on Rocque's map of 1757. The estate was bought by Mr. Sulivan from the Nepean family in 1824, and his daughter, Miss Sulivan, keeps up the garden with the utmost good taste and knowledge of horticulture. The ailanthus, with a trunk 10 feet 4 inches in girth at 4 feet from the ground, is probably one of the finest specimens in England. The one in Fulham Palace garden is exactly the same girth, but does not appear to be so lofty. The liquidamber is also a magnificent tree, and the false acacia is quite as fine as the one in Fulham Palace, and was probably planted at the same time. There are still two cedars left, although the finest was blown down some years ago, and the timber afforded panelling for a large room and many pieces of furniture. Perhaps the most beautiful of the trees is the copper or purple beech. Not only is it very tall and has a massive trunk (14 feet 6 inches at 2 feet from the ground), but the shape is quite perfect, and its branches are furnished evenly all round. There are also good evergreen oaks, elms, chestnuts and Scotch firs. There is a large collection of flowering shrubs, which are in no way affected by the smoky air. Standard magnolias, grandiflora, conspicua and stellata, many varieties of the delightful autumn-flowering plant, the *Hibiscus syriacus*, known to older gardeners as *Althæa frutex*, and recommended under that name by Fairchild in 1722 as suited to London, *Cratægus pyracantha*, *Choysia*, *Pyrus spectabilis*, and many other equally delightful shrubs all appear most flourishing. These, together with herbaceous plants and ornamental trees, well grouped in a garden of good design, with the river flowing at the foot of it, make the grounds of Broom House rank among the most attractive about London.

A few of the gardens, like this one, have succeeded in keeping the real stamp of the country, in spite of the encroachments of the town and the advance of trams and motor omnibuses, but they are every day becoming more scarce. Hampstead and Highgate have many such, and here and there, to the north and on the south of the river, such delightful spots are to be found, although the temptation to cut them up and build small red villas on the sites is very great. Towards the north of London there are many small gardens which are bright and attractive, and without going so far as Hampstead, pleached walks and small but tastefully arranged grounds are

ST. JOHN'S LODGE, REGENT'S PARK

met with. Within Regent's Park there are several charming gardens round the detached villas, which have been already noticed in the chapter on that Park. The two most interesting from a horticultural point of view are St. Katharine's and St. John's Lodges. The fountain in the former is the frontispiece to this volume, and that view says more than any elaborate description. It might be in some far-away Italian garden, so perfectly are the sights and sounds of London obliterated. On a still, hot day, when the fountain drips with a cool sound and there is a shimmering light of summer over the distant trees beyond the terrace, the delusion is perfect. Most of the herbaceous plants which take kindly to London grow in the border—hollyhocks, day lilies, poppies, peonies, pulmoneria and lilies, while there is a large variety of flowering shrubs—ribes, lilacs, buddleias, shumachs and *Aralia spinosa*. The kitchen-garden produces good crops of most of the ordinary vegetables. The garden is arranged with a definite design; there is nothing specially formal, no cut trees or anything associated with some of the formal ideas in England, but there is method in the design; the trees and plants grow as Nature intended them, but they are not stuck about in incongruous disorder and meaningless, distorted lines, as is so often thought necessary, in designing a garden or "improving" a park.

St. John's Lodge has also a well-thought-out garden, some of it of a distinctly formal type. The coloured illustration of it is taken from a part of the garden enclosed with cut privet hedges, with a fountain in the centre, on which stands a statue of St. John the Baptist, by Mr. Johnes. Between the four wide grass walks there are masses of herbaceous plants, backed by rhododendrons, which, as the picture shows, stand out with brilliant colour in summer against the green background. This garden opens into a bowling-green enclosed by cut lime trees, and a cool walk for summer shaded by pleached lime trees. A seductive broad walk bordered with fruit-trees is another feature. This attractive garden has been made within the last eighteen years. The conception of it was due to Lord Bute, and the designing and carrying out to Mr. Schultz. The other side of the house, with a wide terrace and park stretching down toward the water, has no special horticultural feature, but the formal garden is full of charm, and the plants are thriving and trees growing up so fast there is no trace of its newness. It only shows how much can be done where knowledge and good taste are displayed.

St. James's Park is still skirted by garden walls—Stafford, Clarence, and Marlborough Houses, as well as St. James's Palace, though their gardens are hardly as elaborate as those of former years. The garden of that Palace delighted the Sieur de la Serre, who accompanied Marie de Medicis when she came to pay a visit to Henrietta Maria and Charles I. and was lodged in St. James's Palace. After describing the house, " there were, besides," he writes, " two grand gardens with parterres of different figures, bordered on every side by a hedge of box, carefully cultivated by the hands of a skilful gardener; and in order to render the walks on both sides which enclosed it appear more agreeable, all sorts of fine flowers were sowed. . . . The other garden, which was adjoining and of the same extent, had divers walks, some sanded and others grass, but both bordered on each side by an infinity of

IN THE GARDEN, ST. JOHN'S LODGE

fruit-trees, which rendered walking so agreeable that one could never be tired."

The garden of Bridgewater House was a little slice taken off Green Park. On the advice of Fordyce, the Crown in 1795 granted a lease, on certain conditions, to the Duke of Bridgewater and other proprietors near their respective houses, on the ground that it would improve rather than injure the Park. In 1850 the question arose whether the plans Barry had just made for the garden of Bridgewater House infringed the terms of the lease, and Pennethorne, architect to the Office of Works, had to report on the question. It being finally settled that the proposed wall and terrace would not hurt the Park, the alterations were allowed.

Last, but by no means least, either in size or importance, the gardens of Buckingham Palace must be glanced at. The Palace is so modern, when compared with the older Royal residences, that it is easy to forget the history of the forty acres enclosed in the King's private garden, yet they have much historical interest. In the time of James I. a portion of the ground was covered by a mulberry garden, which the King had planted, in pursuance of his scheme to encourage the culture of silkworms, in 1609. That year he spent £935 in levelling the four acres of ground and building a wall round it for the protection of the trees. A few years later most of the enclosure became a tea-garden, while part was occupied by Goring House. There are many references to these famous tea-gardens, called the "Mulberry Garden," in plays and writings of the seventeenth century. Evelyn notes in his "Diary," on 10th April 1654: "My Lady Gerrard treated us at Mulberry Garden, now the only place of refreshment about the town for persons of the best quality to be exceeding cheated at, Cromwell and his partisans having shut up and seized Spring Garden, which till now had been the usual rendezvous for the ladies and gallants at this season."

Goring House stood just where Buckingham Palace does now, and was the residence of George Goring, Earl of Norwich, and of his son, with whom the title became extinct. It was let in 1666, by the last Earl of Norwich, to Lord Arlington, and became known sometimes as Arlington House. It was burnt in 1674, and Evelyn notes in his "Diary" of 21st September : " I went to see the great losse that Lord Arlington had sustained by fire at Goring House, this night consumed to the ground, with exceeding losse of hangings, plate, rare pictures, and cabinets; hardly anything was saved of the best and most princely furniture that any subject had in England. My lord and lady were both absent at the Bath." Buckingham House, which was built in 1703 on the same site for the Duke of Buckingham, must have been very charming. Defoe describes it as " one of the beauties of London, both by reason of its situation and its building. . . . Behind it is a fine garden, a noble terrace (from whence, as well as from the apartments, you have a most delicious prospect), and a little park with a pretty canal." The Duke of Buckingham himself gives a full description of his garden in a letter to a friend, telling him how he passed his time and what were his enjoyments, when he resigned being Privy Seal to Queen Anne (1709). " To the garden," he writes, " we go down from the house by seven steps into a gravel walk that reaches across the garden, with a covered arbour at each end. Another of thirty feet broad leads from the front of the house, and lies between two groves of tall lime trees, planted on a carpet of grass. The outsides of those groves are bordered with tubs of bays and orange trees. At the end of the broad walk you go up to a terrace 400 paces long, with a large semicircle in the middle, from where are beheld the Queen's (Anne's) two parks and a great part of Surrey: then, going down a few steps, you walk on the bank of a canal 600 yards long and 17 broad, with two rows of limes on either side. On one side of this terrace a wall, covered with roses and jessamines, is made low to admit the view of a meadow full of cattle just beneath (no disagreeable object in the midst of a great city), and at each end is a descent into parterres with fountains and waterworks. From the biggest of these parterres we pass into a little square garden, that has a fountain in the middle and two green-houses on the sides . . . below this a kitchen-garden . . . and under the windows . . . of this green-house is a little wilderness full of blackbirds and nightingales." This is truly an entrancing picture of a town garden.

The waterworks, those elaborate fountains then in vogue, were supplied by water pumped up from the Thames into a tank above the kitchen, which held fifty tons of water. Buckingham House was then a red-brick building, consisting of a central square structure, with stone pillars and balustrade along the top, and two wings attached to the main building by a colonnade. It was this style of house when King George III. bought it, originally for a dower-house for Queen Charlotte, instead of Somerset House, where the Queens-Dowager had previously lived. These formal gardens were not suited to the taste of the time, and George IV. had all the garden altered, as well as the house rebuilt by Nash. The whole of the parterres, terraces and fountains and canal were swept away, and most of the lime-trees cut down. A wide lawn and five acres of ornamental water, glades, walks and thickets took their place. When first made the water was severely criticised by a writer of the landscape school, the chief fault he found being that too much was visible at once from the path which encircled it, so that the limits were not well concealed. This seems to have been altered to the satisfaction of later critics. Dennis, writing in 1835, gives a plan in which the path has been made a little distance from the water's edge, and the outline broken by clumps of trees and a promontory, which later on was turned into an island, on which a willow from Napoleon's tomb at St. Helena is said to have been planted, though no old willow now exists. This writer gives great praise to Aiton, who superintended all the execution of the plans. The pavilion in the grounds was added in 1844, and decorated with paintings of scenes from Milton's *Comus* by Eastlake, Maclise, Landseer and other artists, with borders and gilt ornaments by Gruner.

During the last four years his Majesty has had a great deal done to improve the grounds. His appreciation of what is beautiful in gardening has led him to effect several changes, which, while keeping the park-like character of the gardens, have added immensely to their scenic beauty and horticultural interest. The dead and dying trees and others of poor and stunted growth have been removed, giving air and light to those remaining. Several good specimens of plane, lime. elm, beech, ash, ailanthus and hawthorn have thus secured

z

more space to develop. A very large assortment of all
the best flowering shrubs which will flourish in London
have taken the place of worn-out evergreens. The best
of the hollies, arbutus and healthy evergreens have been
encouraged by careful attention. The great object in
laying out the garden originally was naturally to obtain
as much privacy as possible, and the earth taken out of
the lake was formed into a great bank, which was
thickly planted to screen the stables and distant houses.
This bank, which was stiff and formal in appearance,
has now been artistically broken by planting and rock-
work—not merely by a few stones, which would seem
small, unnatural, and out of place, but by bold crags,
over which roses climb, and where gorse, savin and
broom, and countless other suitable plants look perfectly at home. The aspect of the lake is also
greatly enhanced by the substitution of rustic stone
bridges for the iron structures. The water's edge is
well furnished with iris and other water-loving plants
—the finest Marliac lilies brighten its surface—and
the stiff, round island is now varied by striking rocky
promontories and is prettily adorned with broom and
cherries.

The colossal vase by Westmacott, executed as a
memento of the Battle of Waterloo, has lately been
placed on one of the lawns in an amphitheatre of trees.
It stands in front of his Majesty's summer-house,
which is quaint in design, and was brought from the
old Spring Gardens at Whitehall. The views down
the wide glades, with the groups of tall trees, the
bridges, the herbaceous borders, and the wealth of
flowering shrubs, make the garden altogether one of
singular charm considering it is even more truly "in

the midst of a great city" than when the Duke of Buckingham described the same spot nearly 200 years ago.

The Buckingham Palace Gardens show how much judicious planting can do, and how much is lost in many of the parks as well as gardens by not sufficiently considering the decorative value of plants. The old landscape gardeners, in their desire to copy nature and depart from all formality, forgot the horticultural part of their work in their plans for the creation of landscapes. They had not studied the effects which skilful planting will produce, and ignored flowers as a factor in their scenery. They had not got the wealth of genera which the twentieth century possesses, and of which, in many instances, full use is made. But in a review of London Parks and Gardens, it is impossible not to notice effects missed as well as success achieved. The immense advance gardening has made of late years, and the knowledge and wide range of plants, makes it easier to garden now than ever before. The enormous number of trees and flowers now in cultivation leaves a good choice to select from, even among those suitable for the fog-begrimed gardens of London. The carpets of spring flowering bulbs, the masses of brilliant rhododendrons, the groups of choice blossoming trees, which so greatly beautify many of the parks and gardens, are all the result of modern developments. Experience, too, has pointed out the mistakes in landscape gardening, which is for the most part the style followed in London, and it should be easy to avoid the errors of earlier generations. In formal designing, also, the recent introductions and modern taste in flowers should have a marked influence. In all the parks and gardens, public

and private, the chief aim should be to make the best
use of the existing material, to draw upon the vast
resources of horticulture, which have never been so great
as at the present time, and thus to maintain the position
of superiority of London gardens among the cities of
the world.

APPENDIX TO PRIVATE GARDENS

CHARLTON

WING to unavoidable circumstances it was not possible to include Charlton in the foregoing chapter on private gardens, but some account of this place of historic interest is necessary to complete this book. Further from the centres of fashion, on the eastern limits of London, it has not been the scene of such brilliant assemblies as Holland House on the west; yet its early days share that speculative fascination which gathers round the personality of Henry, Prince of Wales, who figures for such a short time on the pages of English history. Only two miles from Greenwich, in the hundred of Blackheath, lies the manor of Charlton, which was bestowed by William the Conqueror on his half-brother, Odo of Bayeux. Later on it passed by gift to the Priory of Bermondsey, and so remained until the Dissolution of the Monasteries, when it became crown land until James I. gave it to Sir Adam Newton, "who built a goodly brave house" thereon. Born in Scotland, Sir Adam had spent much of his life in France, and passing himself off as a priest, had taught Greek at St. Maixant in Poitou. On his return to Scotland in 1600, he was

appointed tutor to Prince Henry, and was in attendance
on him as secretary when the Prince grew up. In 1607
he commenced to build Charlton for him, Inigo Jones
being the architect, and after the Prince's death in 1612,
the King granted Sir Adam the manor, in lieu of payment for the expenses he had incurred in building the
house. The owner of Charlton continued to enjoy royal
favour, became Treasurer of the Household to Prince
Charles, was created a baronet in 1620, and married a
daughter of Sir John Puckering, who had been Keeper of
the Great Seal to Queen Elizabeth. His second son, Sir
Henry Newton, who succeeded him at Charlton, and took
the name of Puckering from estates inherited from his
uncle, was an ardent supporter of Charles I. He sold
the property to Sir William Ducie, Viscount Downe, at
whose death it was again sold. The purchaser, Sir
William Langhorne, was a wealthy East India merchant,
who was, from 1670 to 1677, Governor of Madras. On
his death it passed by entail to his cousin Mrs. Maryon,
and eventually to her great-granddaughter, the wife of
Sir Thomas Spencer Wilson, in whose family Charlton
still remains.

The gardens show traces of all the many owners, and
in spite of the growth of London and its attendant drawbacks, they are still charming. The house stands in
about 150 acres of undulating deer park, with some fine
old trees, an avenue of English elms on the east, and one
of horse-chestnuts, forming the approach on the west.
Perhaps the planting of the tulip tree near the present
lodge was due to John Evelyn, the friend of Sir Henry
Puckering. Evelyn's liking for tulip trees is well known,
and this specimen looks old enough to claim his acquaintance. The two shattered but grand old mulberry trees

probably date from the year 1609, when James I. encouraged all his subjects to plant them, and tradition points to one as the first brought to England. There is an immense horse-chestnut on the lawn, with a wide spread of branches which are rooted in the ground all round, and among the evergreen oaks and other attractive trees in the "Wilderness," a Judas of great age is remarkable. The small house standing near the road which passes the parish church, known as the "Guard House," recalls the time when Prince Henry was living there, and his guard of honour kept watch near the entrance. The stables are just as they were built by Inigo Jones, and the little "Dutch" walled garden which adjoins them on one side is also a pretty relic of those days, and the "Gooseberry Garden" near it is a survival of the same period. A walk overshadowed by tall yew trees stretches across and along the main part of the grounds, and hidden away near its southern end is a delightful rose garden. The beautiful lead fountain in the centre must have been put there by Sir William Langhorne. His initials appear on the leaden tank, and the spray rises from a basin held up by a charming little cupid standing on a pedestal surrounded by swans. The same group appears without the tank in another part of the garden, and there are lead vases and figures, and a cistern dated 1777, which add greatly to the old-world charm which still lingers. Chemical works and sulphurous fumes now work deadly havoc among the old trees, but everything that modern science can recommend is done to preserve them, and young ones planted to keep up the traditions, and bridge over the centuries dividing the present from the days of Prince Henry and his learned and courtly tutor.

LIST OF SOME OF THE WORKS CONSULTED.

(The date does not always refer to the first edition, but to the one consulted.)

Ambulator. A Pocket Companion in a Tour Round London. 1792.
Amusements of Old London. Boulton. 1900.
Argyll, Autobiography of George Douglas, Eighth Duke of. Ed. by
    the Dowager Duchess of Argyll. 1907.

Baker, T. H. Records of Seasons and Prices.
Battersea, All About. H. S. Simmonds. 1882.
Birds in London. W. H. Hudson. 1898.
  ,, of London. H. K. Swann. 1893.
Bloomsbury. Chronicle of Blemundsbury. W. Blott. 1892.
    ,, and St. Giles. George Cluich. 1890.
Bradley, Richard. New Improvements of Planting and Gardening.
    1717.
Burial Grounds, London. Mrs. Basil Holmes. 1896.
Butler, Samuel. Hudibras. Notes by Grey and Nash. 1847.

Calendar of State Papers. 1557, &c. Ed. J. Redington.
Camberwell. Parish of All Saints. T. J. Gaster. 1896.
    ,, Ye Parish of Camerwell. W. H. Blanch. 1875.
Catesby, Mark. Natural History of Carolina. 1731–43.
    ,,   ,, Hortus Europæ Americanus. 1767.
Chelsea. Memoirs of the Botanic Garden. Henry Field. 1820.
  ,,   ,,   ,,   ,, Ed. by R. H. Semple.
    1878.
  ,, An Account of Chelsea Hospital. 1805.
  ,, Historical Notes. Isabella Burt. 1871.
  ,, Hospital. Thomas Faulkener. 1805.
  ,, Thomas Faulkener. 1810.
Cleveland. Character of a London Diurnal. 1647.
  ,, Poems, annotated by J. M. Berden. 1903.
Cole, John. A Pleasant and Profitable Journey to London. 1828.

Commons. A Glance at the Commons and Open Spaces of London. 1867.

Curtis, William. Botanical Magazine. 1787–1906.

,, ,, A Catalogue of the Plants Growing Wild in the Environs of London. 1774.

,, ,, Flora Londinensis. 1777–1828.

Dennis, John. The Landscape Gardener. 1835.

Domesday Book. Ed. 1812.

Draper, W. H. The Morning Walk; or, City Encompass'd. 1751.

Evelyn, John. Diary.

,, ,, Sylva. 1664.

Fairchild, Thomas. The City Gardener. 1722.

Fiennes, Celia. Through England on a Side-Saddle in the Time of William and Mary. Ed. Hon. Mrs. Griffiths. 1888.

Foreign Visitors to England. Smith. 1889.

Fulham, Old and New. C. J. Feret. 1900.

,, and Hammersmith. Faulkener. 1813.

Gardeners' Magazine. Conducted by J. C. Loudon. 1826–43.

Gardening. History of, in England. Alicia Amherst. 1896.

Gerard. Herbal. 1597.

,, ,, Ed. by T. Johnson. 1633.

,, Catalogus. 1599.

Greenwich. W. Howarth. 1886.

,, and Blackheath. Half Holiday Hand-book Series. 1881.

,, Park: Its History and Associations. Angus D. Webster. 1902.

,, The Palace and Hospital. A. G. K. L'Estrange. 1886.

Grosley. A Tour to London. 1765.

Hackney. Magazine and Parish Reformer. 1833–38.

,, Collecteanea Geographica, &c. 1842.

,, History and Antiquities of. William Robinson. 1842.

Hawthorne, Nathaniel. Passages from English Note-Books. 1870.

Hazlitt, W. C. Gleanings in Old Garden Literature. 1887.

Highgate, History of. Frederick Prickett. 1842.

Hook, Dean of Chichester. Lives of the Archbishops of Canterbury. 1875.

Hyde Park, from Domesday to Date. J. Ashton. 1900.

Index Kewensis. 1893, &c.
Inns of Court. Inner Temple Records. F. A. Inderwick. 1896.
    ,,   ,,   Inner and Middle Temple. H. H. L. Bellot. 1902.
    ,,   ,,   Lincoln's Inn. Douthwaite. 1886.
    ,,   ,,   Gray's Inn. Douthwaite. 1886.
    ,,   ,,   and Chancery. W. J. Loftée. Illustrations by Herbert Railton. 1893.
Islington. History of the Parish of St. Mary. S. Lewis. 1842.
Issue Rolls. James I., &c.

Lamb, Charles, Life of. E. V. Lucas. 1905.
Lambeth, History of. Ducarel. 1785.
    ,,   ,,   Thomas Allen. 1828.
    ,,   Palace and its Associations. J. C. Browne. 1883.
Laud, Archbishop's, Diary.
Loimographia: An Account of the Great Plague. W. Boghurst. 1894.
London, Ancient and Modern, from a Sanitary Point of View. G. V. Poore. 1889.
    ,,   Birds and Insects. T. D. Pigott. 1892.
    ,,   Botanic Gardens. Pierre E. F. Perrédès. Pub. by Wellcome Chemical Research Laboratories. No. 62.
    ,,   Bygone. F. Ross. 1892.
    ,,   City Suburbs as they are To-Day. 1893.
    ,,   City: Its History, &c. W. J. Loftie.
    ,,   Curiosities of. Timbs. 1868.
    ,,   Environs of. Daniel Lysons. 1790–96.
    ,,   Familiar. J. C. L'Estrange. 1890.
    ,,   Fascination of. Series ed. by Sir W. Besant.
    ,,   Flora. Alexander Irvine. 1838.
    ,,   Garland. W. E. Henley. 1895.
    ,,   Greater. E. Walford. 1893–95.
    ,,   Hand-book of. Peter Cunningham. 1850.
    ,,   Highways and Byways in. Mrs. E. T. Cook. 1903.
    ,,   History of. Noorthouck. 1773.
    ,,   ,,   William Maitland. 1756.
    ,,   ,,   Plantagenet, Tudor Times, &c. Sir W. Besant.
    ,,   Illustrata. Wilkinson.
    ,,   Its Neighbourhood, &c. Hughson David. 1805–9.

London, Journey to.   John Cole.   1825.
,, Knight, Charles.   Revised by E. Walford.
,, Life Seen Through German Eyes.   Brand.   1887.
,, Memories.   C. W. Heckethorne.   1900.
,, Our Rambles in Old.   E. S. M. Smith.   1895.
,, Pageant of.   Richard Davey.   1906.
,, Past and Present.   H. B. Wheatley and P. Cunningham.   1891.
,, Pleasure Gardens of the Eighteenth Century.   W. Wroth.
         1896.
,, Redivivium.   James Peter Malcolm.   1807.
,, Reliques of Old London and Suburbs.   H. B. Wheatley.
         1896.
,, Round About.   W. J. Loftée.   1893.
,, Signs and Inscriptions.   Wheatley and Philip Norman.   1893.
,, Some Account of.   Thomas Pennant.   1793.
,, Soul of.   F. H. Madox Heuffer.   1905.
,, Story of.   H. B. Wheatley.   1904.
,, Survey of.   (London County Council.)   C. R. Ashbee.
         1900.
,,   ,,   Stowe.   Several Editions.   1598, 1633, &c.
,, of To-Day.   C. E. Pascoe.   1885.
,, Town.   Marcus Fall.   1880.
,, Vanished and Vanishing.   P. Norman.   1905.
,, Vestiges of Old.   Archer J. Wykeham.   1851.
,, Walks Through.   Hughson David.   1817.
,,   ,,   In.   Augustus Hare.   1901.
Londres et Les Anglais en 1771.   Join Lambert.   1890.
London, G., and H. Wise.   Complete Gardener.   1701.
Loudon, J. C.   Arboretum.   1838.
,,   Encyclopædia of Gardening.   1822.
,,   ,,   of Plants.   1838.
,,   Gardeners' Magazine.
,,   Laying Out, &c., of Cemetries.   1843.

Magalotti.   Travels of Cosmo III., Grand Duke of Tuscany, through
      England, 1669.   1821.
Maitland, William.   History and Survey of London.   1756.
Marylebone, Random Sketches in.   F. H. Hallam.   1885.
,,   and St. Pancras.   G. Cluich.   1890.
Mayfair and Belgravia.   G. Cluich.   1892.

Miller, Philip. Gardeners' and Florists' Dictionary. 1724.
  „   „   „   Dictionary. 1759.
Mirabeau. Letters during his Residence in England. 1832.
Misson, H. Memoirs and Observations in his Travels over England.
  Translated by Mr. Ozell. 1719.
Montagu, Letters of Mrs. E. 1809–13.
  „   Mrs. E. By J. Doran. 1873.
Montgomerie, James. Chimney Sweepers' Friend. 1824.
Municipal History, Bibliography of. Cross. 1897.

Nichol. Progress of Queen Elizabeth.
Nisbet, J. British Forest Trees. 1895.
Norden. Notes on his Map of London, 1593. H. B. Wheatley.
  1877.

Open Lands, Inclosure and Preservation of. Sir Robert Hunter. Re-
  print. 1897.

Parliamentary Reports—
  Committee on the Public Parks, &c. 1887.
    „   „   Best Means of Preserving . . . Use of Forests,
    Commons, &c. 1865.
  Other Reports : see Catalogue of Parliamentary Papers, 1801–1908.
    P. and S. King & Co.
  Plan of Improvements proposed opposite Buckingham Palace.
    1850.
  Return of the Outlay on Battersea Park. 1856.
  Select Committee on Open Spaces. 1865.
Parks, Gardens, &c., of London. Edward Kemp. 1851.
  „   Hyde Park, from Domesday to Date. J. Ashton. 1896.
  „   Municipal, and Gardens. Lieut.-Col. J. G. Sexby. 1905.
  „   and Pleasure Grounds. C. H. J. Smith. 1852.
  „   „ Open Spaces. London County Council Sixpenny Guide.
        1906.
  „   „   „   and Thoroughfares. A. M'Kenzie. 1869.
  „   Royal, and Gardens. N. Cole. 1877.
  „   Story of the London. Jacob Larwood. 1872.
Parkinson, John. Paradise in Sole. 1629.
Pepys, Samuel. Diary.
Piccadilly and Pall Mall, Round About. H. B. Wheatley. 1817.

Philips, Henry. Sylva Florifera. 1823. Flora Historica, &c.
Phillips, Sir Richard. Morning Walk to Kew. 1817.
Pulteney, Richard. History of the Progress of Botany in England. 1790.
Pyne, Wm. H. History of the Royal Residences. 1819.

Regent's Park. Some Account of the Improvements. 1814.
„   „   „   „   „   John White. 1815.
„   „   Literary Pocket Book. 1823.
„   „   Picturesque Guide to. 1829.
Repton. Landscape Gardening. Ed. J. C. Loudon. 1840.

St. Botolph, Aldgate. A. G. B. Atkinson. 1898.
St. James's Square. Dasent. 1895.
Selby, P. J. British Forest Trees. 1841.
Shipton, Mother. Life and Death of. 1687.
„   „   Prophecies. Ed. E. Pearson. 1871.
„   „   „   C. Hindley. 1877.
Soho, Two Centuries of. J. H. Cardwell. 1898.
„   and its Associations. E. F. Rimbault. 1895.
Sorbière, Samuel de. A Journey to London. [William King.] 1698.
„   „   A Voyage to England. 1709.
„   „   Journey to London. 1832.
„   „   Reponse aux Faussetés . . . dans la relation du Voyage en Angleterre. 1675.
Stepney. Two Centuries of History. W. H. Frere. 1892.
Stowe. Survey of London. 1598.
„   Munday's Edition. 1633.
„   Strype's Edition. 1720.
Suburban Reliques of Old Londons. H. B. Wheatley. Drawn by T. R. Way. 1715.
Switzer, Stephen. Nobleman, Gentleman, and Gardener's Recreation. 1715.

Tradescant, John. Museum Tradescantianum. 1656.
Trinity Hospital, Mile End Road. C. R. Ashbee. 1896.

Westminster, Antiquities of. John T. Smith. 1807.
„   Abbey. W. J. Loftie. 1890.

Westminster Abbey. Dean Stanley.

      ,,   ,,   MSS. Records.

      ,,   ,,   Richard Widmore. 1751.

      ,,   Memorials of the City, St. Peter's College, &c. Rev. MacKenzie E. C. Walcott. 1849.

Wheatley. Observations on Modern Gardening. 1793.

Whitten, W. London in Song. 1898.

Wren, Christopher. Parentalia. 1750.

HYDE PARK AND KENSINGTON GARDENS

LIST OF TREES AND SHRUBS

N.B.—Those marked thus * are not in existence at the present time. A small number proved unsuitable for London, and others have been removed from the plantations for various reasons.

Acer campestre.
  ,, circinatum.
  ,, creticum.
  ,, dasycarpum.
  ,, macrophyllum.
  ,, Negundo.
  ,,   ,, foliis variegatis.
  ,, palmatum.
  ,, platanoides.
  ,,   ,, Reitenbachii.
  ,,   ,, Schwedleri.
  ,, Pseudo-platanus.
  ,,   ,,   ,, foliis variegatis.
  ,,   ,,   ,, purpureum.
  ,, rubrum.
  ,, saccharum.
  ,, saccharum nigrum.
  ,, tartaricum.
Æsculus Hippocastanum.
  ,,   ,, laciniata.
  ,,   ,, rubicunda.
Ailantus glandulosa.
Alnus barbata.
  ,, cordifolia.
  ,, glutinosa.
  ,,   ,, incisa.
  ,,   ,, laciniata.
  ,,   ,, quercifolia.
Amorpha fruticosa.
Amygdalus (Prunus) communis.
  ,,   ,,   ,, amara.

Amygdalus communis macrocarpa.
  ,, nana.
Amelanchier canadensis.
  ,, vulgaris.
Aralia chinensis.
  ,, spinosa.
Arbutus Andrachne.
  ,, Unedo.
  ,,   ,, rubra.
Aristolochia Sipho.
Armeniaca (Prunus) sibirica.
Artemisia arborescens.
Asimina triloba.
Aucuba japonica.
  ,,   ,, maculata.
  ,,   ,, viridis.
Azalea (Rhododendron) sinense.
  ,, pontica.
  ,, nudiflorum.

Berberis Aquifolium.
  ,, Darwinii.
  ,, Fortunei.
* ,, japonica.
  ,, repens.
  ,, stenophylla.
  ,, vulgaris.
  ,,   ,, foliis purpureis.
Betula alba.
  ,,   ,, pendula.

Betula fruticosa.
„ lenta.
„ nana.
„ nigra.
„ populifolia.
„ urticifolia.
Buxus balearica.
„ caucasica.
„ sempervirens arborescens.
„ „ aureo-marginita.

Caragana arborescens.
„ Chamluga.
„ frutescens.
„ spinosa.
Carpinus betulus.
Carya amara.
Caryopteris Mastacanthus.
Castanea sativa.
Catalpa bignonioides.
*Cedrus Deodora.
„ Libani.
Cerasus. See Prunus.
Cercis Siliquastrum.
Cistus florentinus.
„ ladaniferus.
„ monspeliensis.
Clematis Flammula.
„ Jackmani.
„ montana.
„ Vitalba.
Celtis Tournefortii.
Clerodendron trichotomum.
Colutea arborescens.
Cornus alba.
„ „ Spæthii.
„ Mas.
„ „ aurea elegantissima.
„ „ variegata.
„ sanguinea.
„ stolonifera.
Coronilla Emerus.
Coryllus Avellana.
„ maxima atropurpurea.
Cotoneaster acuminata.
„ bacillaris.

Cotoneaster frigida.
„ horizontalis.
„ microphylla.
„ Nummularia.
„ Simmonsii.
Cratægus altaica.
„ Azarolus.
„ coccinea.
„ cordata.
„ „ accrifolia.
„ „ maxima.
„ Crus-galli.
„ „ ovalifolium.
„ „ pyracanthafolia.
„ „ splendens.
„ dippeliana.
„ heterophylla.
„ macrantha.
„ nigra.
„ orientalis.
„ Oxyacantha.
„ „ aurea.
„ „ eriocarpa.
„ „ flexuosa.
„ „ flore pleno albo.
„ „ flore pleno coccineo.
„ „ flore pleno puniceo.
„ „ flore pleno roseo.
„ „ flore pleno rubro.
„ „ flore roseo.
„ „ laciniata.
„ „ pendula.
„ „ præcox.
„ „ quercifolia.
„ „ stricta.
„ punctata.
„ „ brevispina.
„ „ xanthocarpa.
„ pyracantha.
„ „ Lalandi.
„ siniaca.
„ spathulata.
„ tanacetifolia.
*Cupressus Lawsoniana.
* „ Nootkatensis.
* „ sempervirens.

2 A

Cydonia japonica.
„ Maulei.
„ vulgaris lusitanica.
„ „ maliformis.
Cytisus albus.
„ alpinus.
„ nigricans.
„ præcox.
„ racemosus.
„ scoparius.
„ sessilifolius.
„ tinctoria.

Daphne Mezereum.
„ pontica.
Diospyros Lotus.
„ virginiana.
Diplopappus chrysophylla.
Deutzia crenata.
„ „ flore pleno.
„ „ gracilis.
„ scabra.

Elæagnus angustifolia.
„ argentea.
Euonymus europæus.
„ „ fructo albo.
„ japonicus.
„ „ argenteus.
„ „ aureo-variegatus.
„ „ radicans.
„ „ „ foliis pictis.
„ latifolius.

Fagus sylvatica.
„ „ cuprea.
„ „ pendula.
„ „ purpurea.
„ „ „ pendula.
Fatsia japonica.
Ficus Carica.
Fontanesia phillyræoides.
Forsythia intermedia.
„ suspensa.
„ viridissima.
Fraxinus americana cinerea.

Fraxinus americana elliptica.
„ „ juglandifolia.
„ excelsior.
„ „ angustifolia.
„ „ aurea.
„ „ heterophylla.
„ „ pendula.
„ Ornus.
„ „ angustifolia.
„ parvifolia.

Genista hispanica.
Gleditschia triacanthos.
„ sinensis.
„ „ nana.
Gymnocladus canadensis.

Halesia diptera.
„ tetraptera.
Halimodendron argenteum.
Hamamelis virginica.
Hedera Helix.
„ „ arborescens.
„ „ caenwoodiana.
„ „ canariensis.
„ „ „ arborescens.
„ „ chrysocarpa.
„ „ colchica.
„ „ dentata.
„ „ digitata.
„ „ lucida.
„ „ maderensis variegata.
„ „ minima.
„ „ taurica.
„ „ variegata.
Hibiscus syriacus—
and numerous garden varieties.
Hippophæ rhamnoides.
„ salicifolia.
Hydrangea hortensia.
„ paniculata grandiflora.
Hypericum calycinum.
„ elatum.
„ hircinum.
„ patulum.

Ilex Aquifolium.
  ,,   ,,   albo-picta.
  ,,   ,,   altaclerense.
  ,,   ,,   angustifolia.
  ,,   ,,   ,, variegata.
  ,,   ,,   argentea variegata.
  ,,   ,,   argentea marginata.
  ,,   ,,   aureo-picta.
  ,,   ,,   aureo-regina.
  ,,   ,,   balearica.
  ,,   ,,   camelliæfolia.
  ,,   ,,   ferox.
  ,,   ,,   ,, argentea.
  ,,   ,,   ,, aurea.
  ,,   ,,   fructo luteo.
  ,,   ,,   heterophylla.
  ,,   ,,   Hodginsii.
  ,,   ,,   latispina.
  ,,   ,,   laurifolia.
  ,,   ,,   myrtifolia.
  ,,   ,,   recurva.
  ,,   ,,   scotica.
  ,,   ,,   Shepherdii.
  ,,   ,,   Watereriana.
  ,, dipyrena
  ,, latifolia.
  ,, opaca.

Jasminum fruticans.
  ,,   humile.
  ,,   nudiflorum.
  ,,   officinale.
*Juniperus chinensis.
* ,,   communis.
* ,,   nana.
  ,,   Sabina tamariscifolia.
  ,,   ,, procumbens.
* ,,   virginiana.
Juglans cinerea.
  ,, nigra.
  ,, regia.

Kerria japonica.
Kœlreuteria paniculata.

Laburnum alpinum.
  ,,   vulgare.

Laburnum vulgare quercifolium.
  ,,   ,, Watereri.
Laurus nobilis.
Leycesteria formosa.
Ligustrum Ibota.
  ,,   japonicum.
  ,,   lucidum.
  ,,   ovalifolium.
  ,,   ,, foliis aureis.
  ,,   Quihoui.
  ,,   vulgare.
Liquidamber styraciflua.
Liriodendron tulipifera.
Lonicera Caprifolium.
  ,,   flexuosa.
  ,,   involucrata.
  ,,   Periclymenum.
Lycium chinense.
* ,,   hamilifolium.

Magnolia acuminata.
  ,,   conspicua.
  ,,   ,, Soulangeana.
  ,,   grandiflora.
  ,,   stellata.
Morus alba.
  ,,   ,, pendula.
  ,, nigra.

Osmanthus Aquifolium ilicifolius.

Pavia (Æsculus) flava.
  ,,   ,, purpurascens.
  ,,   glabra arguta.
  ,,   humulis.
  ,,   neglecta.
  ,,   parvifolia.
  ,,   rubra.
Philadelphus coronarius.
  ,,   ,, tomentosus.
  ,,   floribundus.
  ,,   ,, verrucosus.
  ,,   Gordonianus.
  ,,   grandiflorus floribundus.
  ,,   hirsutus.
  ,,   inodorus.

Philadelphus Lemoinei.
Phillyrea angustifolium.
  ,,   decora.
  ,,   latifolia.
Photinia serrulata.
*Pinus cembra.
* ,, insignis.
* ,, Laricio.
  ., sylvestris.
*Planera aquatica.
* ,, Richardi.
Platanus accrifolia.
Populus alba.
  ,,   ,, pyramidalis (bolleana).
  ,, balsamifera.
  ,, canescens.
  ,, deltoidea.
  ,,   ,, aurea.
  ,, macrophylla.
  ,, nigra.
  ,,   ,, betulæfolia.
  ,,   ,, pyramidalis.
  ,, tremula.
  ,,   ,, pendula.
Prunus including Cerasus and persica.
  ,, persica camelliæflora.
  ,,   ,, flore roseo pleno.
  ,,   ,,   ,, alba pleno.
  ,,   ,, dianthiflora pleno.
  ,, Avium.
  ,,   ,, flore pleno.
  ,,   ,, pendula.
  ,, cerasifera.
  ,,   ,, atropurpureum.
  ,, communis.
  ,, (Cerasus) acida semperflorens.
  ,, japonicas flore roseo pleno.
  ,, pseudo-cerasus.
  ,, (Padus) Mahaleb.
  ,,   ,, pendula.
  ,, Padus.
  ,, rotundifolia.
  ,, serotina.
  ,, (Laurocerasus) caucasica.
  ,,   ,, colchica.
  ,,   ., Laurocerasus.

Prunus (Laurocerasus) lusitanica.
  ,, serrulata.
  ,, spinosa.
  ,, triloba.
  ,, Watereri.
Ptelea trifoliata.
Pterocarya caucasica.
Pyrus Aria.
  ,,   ,, salicifolia.
  ,,   ,, undulata.
  ,, amygdaliformis.
  ,, arbutifolia.
  ,, Aucuparia.
  ,, auricularis.
  ,, baccata.
  ,, communis.
  ,, floribunda.
  ,, hybrida.
  ,, intermedia.
  ,, lanata.
  ,, malus astracanica.
  ,, nivalis.
  ,, pinnatifida.
  ,, rivularis.
  ,, spectabilis.

Quercus Ægilops.
  ,, cerris.
  ,,   ,, cana-major.
  ,,   ,, cana-minor.
  ,,   ,, fulhamensis.
  ,, coccinea.
  ,, fastigiata.
  ,, filicifolia.
  ,, Ilex.
  ,,   ,, Gramuntia.
  ,, lucombeana.
  ,, palustris.
  ,, pedunculata.
  ,,   ,, fastigiata.
  ,, rubra.
  ,,   ,, longifolia.
  ,, Suber.

Rhamnus Alaternus maculata.
  ,, alpina.

Rhamnus cathartica.
  „   Frangula.
  „   infectoria.
Rhododendron Cunninghami.
  „   dauricum.
  „   hybrids in variety.
  „   ponticum.
  „   præcox.
Rhus canadensis.
  „   copallina.
  „   cotinus.
  „   glabra laciniata.
  „   typhina.
  „   „   frutescens.
Ribes alpinum.
  „   „   pumilum.
  .,   aureum.
  „   „   præcox.
  „   Diacantha.
  „   nigrum variegatum.
  „   Sanguineum.
  „   „   albidum.
Robinia hispida.
  „   inermis.
  „   Pseudacacia.
  „   „   angustifolium.
  „   „   bessoniana.
  „   „   Decaisneana.
  „   „   dubea.
  „   „   elegans.
  „   „   fastigiata.
  „   „   heterophylla.
  „   „   inermis.
  „   „   monophylla.
  „   „   semperflorens.
  „   „   tortuosa.
  „   viscosa.
Rosa arvensis.
  „   Banksiæ.
  „   canina.
  „   damascena.
  „   gallica centifolia.
  „   „   muscosa.
  „   indica.
  „   multiflora.
  „   noisettiana.

Rosa rubiginosa.
  „   rugosa.
  „   „   flore pleno.
  „   wichuraiana.
  „   hybrids in variety.
Rosmarinus officinalis.
Rubus fruticosus.
  „   „   albo-pleno.
  „   „   rubra-pleno.
  „   laciniatus.
  „   nutkanus.
Ruscus aculeatus.

Salisburia (Ginkgo) adiantifolia.
Salix alba.
  „   babylonica.
  „   Caprea.
  „   daphnoides.
  „   rosmarinifolia.
  „   viminalis.
Sambucus nigra.
  „   „   laciniata.
  „   „   foliis aureis.
  „   racemosa.
  „   „   plumosa.
  „   „   „   aurea.
Skimmia Fortunei.
  „   japonica.
Spartium junceum.
Smilax aspera.
  „   glauca.
  „   rotundifolia.
Sophora japonica.
Spiræa bullata.
  „   canescens.
* „   cantoniensis.
* „   chamædrifolia.
* „   discolor.
* „   japonica.
  „   „   Bumalda.
  „   prunifolia flore pleno.
  „   salicifolia.
  „   sorbifolia.
  „   Thunbergii.
Symphoricarpus orbiculatus.
  „   racemosus.

Syringa Emodi.
,,   Josikæa.
,,   persica.
,,   ,,   alba.
,,   vulgaris.
And many garden varieties.

Taxodium distichum.
Taxus baccata.
,,   ,,   adpressa.
,,   ,,   ,,   aurea.
,,   ,,   Dovastoni.
,,   ,,   fastigiata.
,,   ,,   fructo luteo.
,,   canadensis.
,,   cuspidata.
*Thuja dolobrata.
*   ,,   japonica.
,,   occidentalis.
,,   orientalis.
,,   ,,   aureo-variegata.
*   ,,   plicata.
Tilia americana.
,,   argentea.
,,   cordata.
,,   dasystyla.
,,   petiolaris.
,,   platyphyllus asplenifolia.
,,   vulgaris.

Ulex europæus.
,,   ,,   flore pleno.
,,   nanus.
Ulmus americanus.
,,   ,,   pendula.
,,   campestris.
,,   ,,   Louis van Houtte.

Ulmus campestris sarniensis.
,,   ,,   Wheatleyi.
,,   glabra.
,,   ,,   cornubiensis.
,,   ,,   stricta.
,,   montana.
,,   ,,   atropurpureum.
,,   ,,   fastigiata aurea.
,,   ,,   pendula.
,,   ,,   vegeta.
,,   pedunculata.

Veronica cupressoides.
,,   Traversii.
Viburnum dentatum.
,,   Lantana.
,,   Lentago.
,,   Opulus.
,,   ,,   sterile.
,,   Tinus.
,,   ,,   hirtum.
,,   plicatum.

Weigela (Diervilla) florida.
,,   hybrida.
,,   Looymansi aurea.
Wistaria chinensis.
,,   multijuga.

*Xanthorrhiza apiifolia.

Yucca angustifolia.
,,   filamentosa.
,,   ,,   flaccida.
,,   gloriosa.
,,   recurvifolia.

EXAMPLES OF PLANTING FLOWER-BEDS
IN HYDE PARK IN 1905-6

BED 1.

1. Autumn planting for spring flowers :—Hyacinths, margin of Saxifrage. Day Lily, thinly planted, for bright green foliage growing up with and above the Hyacinths.

2. Spring planting for early summer flowers :—Pansies for margin 18 inches wide, the centre of bed planted with Ragged Robin.

3. Summer planting for later summer and autumn display :—Large plants of Calceolaria Burbidgeii 8 feet high, Cassia corymbosa 6 feet high, Heliotrope 6 feet to 7 feet high, finishing off with Nicotiana affinis and sylvestris, Lantana Drap d'Or with Lilium longiflorum interspersed.

BED 2.

1. Autumn planting for spring flowers :—Tulips, margin of Saxifrage. Iris germanica for foliage planted thinly with bulbs.

2. Delphiniums, deep blue, 18 inch margin of yellow Pansies.

3. Broad margin of Dell's dark Beet, remainder of bed well planted with Cannas, Alphonse Bouvier, and Flambeau, brilliant crimson flowers.

BED 3.

1. Autumn planting for spring flowers :—Narcissus Emperor with a 6 to 1 mixture of Hyacinth King of the Blues, margin of Saxifrage.

2. Broad margin of Pansies, remainder of bed filled with Erigeron speciosum.

3. Large plants in pots of Ivy-leaved Pelargonium Madame Crousse, 6 feet high, placed 5 feet apart. Margin and intermediate spaces planted with dwarf plants of a deeper coloured Ivy-leaved Pelargonium.

BED 4.

1. Autumn planting for spring flowers :—Hyacinths Czar Peter, light-blue, Gigantea blush, margin of Saxifrage.

2. Dictamnus in two colours, about 2 feet apart, ground of bed Anemone coronaria margined with Saxifraga Camposii.

3. Gymnothrix latifolia, Kochia scoparia tricophylla interspersed with Acalypha musaica.

375

BED 5.

1. Dark Wallflowers with margin of Gardiner's Garter (Phalaris).
2. Pelargonium Achievement 4 feet high and 4 feet apart, centre of bed and margin planted with dwarf plants of same variety.
3. Celosia pyramidalis crimson and gold, with some crimson Cockscombs intermixed, the remaining portion of bed thickly planted to the margin with Leucophytum Brownii.

BED 6.

1. Autumn planting for spring flowers:—Hyacinth Grande Maitre, blue.
2. An interesting combination of the following flowers in rotation, fresh ones being introduced as others faded :—Linum perenne, Ixias, Sparaxis, and Calochortus, in variety, Oxalis rosea, Camassia esculenta, Lychnis Viscaria, Crassula coccinea, Balsams with double pink blooms. The setting for these flowers was a variegated grass. A good effect was the result for many weeks.
3. For the remainder of the season this bed was filled with a succession of Lilium speciosum roseum on a green ground, with a margin of Agathea cœlestis.

ERRATA

Page 16, line 24, *for* 'Sir John Sloane' *read* 'Sir Hans Sloane.'

,, 42, ,, 4, *for* 'places' *read* 'plans.'

,, 77, ,, 15, *for* 'Quintinge' *read* 'Quintinye.'

,, 241, ,, 7, *for* 'battle of Alma' *read* 'battle of the Alma.'

INDEX

THE END

Printed by BALLANTYNE, HANSON & Co.
Edinburgh & London

Printed in the United States
By Bookmasters